Carl von Linne

Institutions of Entomology

Being a Translation of Linnaeus's Ordines ET Genera Insectorum

Carl von Linne

Institutions of Entomology

Being a Translation of Linnaeus's Ordines ET Genera Insectorum

ISBN/EAN: 9783337813581

Printed in Europe, USA, Canada, Australia, Japan

Cover: Foto ©berggeist007 / pixelio.de

More available books at **www.hansebooks.com**

INSTITUTIONS

OF

ENTOMOLOGY:

Being a Translation of

LINNÆUS'S ORDINES ET GENERA INSECTORUM;

OR,

SYSTEMATIC ARRANGEMENT OF INSECTS.

COLLATED WITH

The DIFFERENT SYSTEMS of

GEOFFROY, SCHÆFFER and SCOPOLI;

TOGETHER WITH

OBSERVATIONS of the TRANSLATOR.

By THOMAS PATTINSON YEATS.

LONDON:
Printed for R. HORSFIELD, No. 22, Ludgate street.
MDCCLXXIII.

PREFACE.

SOME Friends, for whose judgment I entertain the highest deference, having repeatedly requested me to add an engraving of each Genus of Insects to the following Work, I think myself under an obligation to inform them of the reasons which prevented my complying with their desire. The extraordinary expence which would have attended such engravings must necessarily have enhanced the price of the Work, and defeated its principal design, by preventing a number of such persons as have most occasion for it from purchasing it; this expence would have been so much the greater, as it would not have sufficed to have figured one insect of each Linnæan Genus, but would have been absolutely necessary to have given one, at least, of each family or section of such genera as contain insects differing much from one another in their external appearance: add to this, that I could have done little more than copy the excellent figures of Geoffroy and Schæffer, the first of which are to be purchased at as easy a rate, perhaps, as I could have afforded them, with the advantage of adding another useful work to the purchaser's library. As this book, however, is written in a language not universally understood, such persons as think new engravings absolutely necessary to the well understanding Linnæus's System, will, I hope, soon have an opportunity of procuring them; for, sensible that nothing is so conducive to the perfection of entomological science as the knowledge of the species, I intend communicating to the Public descriptions and figures, coloured after nature, of a large number of very rare insects, from different parts of the world, amongst which will be contained some not met with in any collection but my own, and not described by any author whatever. In this Work, I shall endeavour to insert one specimen, at least, of every Linnæan Genus, with the parts from which the generical characters are taken, delineated in such a manner as

to

to obviate every difficulty, making, at the same time, such alterations in the system of that author, as insects, with characters unknown to him at the time of his framing it, shall render absolutely necessary.

With respect to the following sheets, if, on the one hand, I have not answered the expectations of my friends, I flatter myself that I have exceeded them on the other, by extending my plan considerably beyond the original design; for I have not only taken all the pains in my power to render the meaning of Linnæus as plain as possible; but considering that it would be an advantage to beginners to be acquainted with the systems of some other authors, each of which has his separate admirers, and has made considerable alterations in that of Linnæus, I willingly undertook to collate and compare those different Systems, and explain the reasons which induced their authors to differ from their common Master. The most distinguished among these are Geoffroy, Scopoli and Schæffer; the first of whom, in his *His-*

toire

toire Abregeé des Infectes, publifhed at Paris in 1764, has befides changing the orders, or firft grand divifions, of the Linnæan Syftem, formed from the different families of Linnæan genera, many new genera, fome of them very judicioufly, others, perhaps, without fufficient grounds. It may, however, be faid, in defence of his frequent divifions of the Linnæan genera, that, as his Syftem was a partial one, confined to the infects of a fmall diftrict, he could not take notice, in his Work, of thofe, (as I may call them) intermediate infects, which connect the feveral families, and prove them to belong to the fame genus, fuch infects being frequently exotic.

Scopoli, in his *Entomologia Carniolica*, publifhed at Vienna in 1763, has made few alterations in the Linnæan Syftem; but thofe feem every one to be well founded, and his fpecific characters equal thofe of Linnæus. Schæffer, in his *Elementa Entomologiæ*, printed at Ratifbon, in 1766, has followed Geoffroy with very few and inconfiderable variations; but his figures convey

convey a pretty good idea of his genera, though they cannot be pronounced superior to those of that author. I should have been glad to have given some account of the System of Poda, a Jesuit, a work much praised by Scopoli, which alone is sufficient to convey an advantageous idea of it, but have not been able to procure it, nor learn how or in what he differs from Linnæus.

The Reader will find, that I have not only explained the circumstances from which the above-mentioned Authors have taken their classical and generical distinctions, but likewise the more minute ones, which induced them to form their *genera* into *sections* or *families*. By these means the beginner, instead of contenting himself with attending to a few of the more striking characters, will be led to the consideration of every part of the insect; and as the best method of becoming acquainted with those characters is the comparing of insects known to belong to a certain particular genus, with the description given of that genus, I have taken care

(when

when I could learn it) to apply to each could it moſt familiar Engliſh name, by which any ſpecies belonging to it is known.

If this Eſſay ſhall conduce to the rendering ſo rational an entertainment as the contemplation and ſtudy of the works of Nature more univerſal, or more pleaſing, I ſhall think my trouble more than repaid, and wiſh it no other ſucceſs than that its defects may induce ſome more able Entomologiſt to favour the Public with one more perfect.

CORRECTIONS and ADDITIONS.

Page 25, l. 1, inſtead of *Chermes and Coccus*, read, *the Chermes of Linnæus*. Page 26, l. 6, for *ginus* read *genus*. Page 27, l. 22, for *Anthers*, read *Authors*. Page 48, l. 17, for *Crioceris*, read *Crioceris*. Page 49, l. 3, for *in general*, read *for the moſt part*. Page 99, l. 13, after the word *crotchet* add, *and a ſetaceous, lateral, articulated finger; being &c.* Page 111, l. 3, for *containing*, read *conſiſting*. Page 112, l. 1, for *beliſirm*, read *chenſirm*. Page 122, l. 18, after the word *they*, add *ſo*. Page 123, l. 3, 4, omit the words *like Chermes*. Page 127, l. 5, for *generically*, read *generally*. Page 136, l. 3, 4. for *ſomewhat reſembling*, read *and ſomewhat reſemble*. Pag. 154, l. 14, for *larva*, read *larvæ* Id. l. 22, for *Phryganea*, read *Phryganiæ*. Page 165, l. 5, omit the word *and*. Page 181, l. 8, add after the word of *generical*, *or*. Id. l. 9, for *names*, read *name*. Page 208, l. antipenult. for *their*, read *the*. Page 209, for *midſile*, read *middle*. Page 213, l. 16, for *Volumcella*, read *Volucella*. Line 24, id. id. Page 229, l. penult. for *chryſalis*, read *chryſalids*. Page 244, l. 3, for *larva*, read *larvæ*.

INSECTS.

Linnæus Syst. Nat. Vol. I. P. 2, P. 533.

Properties peculiar to INSECTS, and the Characters by which they are distinguished from the other Classes of the Animal Kingdom.

INSECTS are small animals, having many feet, and breathing through pores arranged along their sides. Their skin (with which they are covered as with a coat of mail) is of a hard or boney consistence. They are furnished with moveable antennæ, growing from the head, and which seem to be endued with an exquisite sense of feeling.

The *body* of these animals is composed of a *head*, a *trunk*, an *abdomen*, and *limbs*.

The *head* is for the most part distinct from the *trunk*, being attached to that part by a kind of articulation or joint. It is furnished with *eyes*, *antennæ*, and, in general, with a *mouth*, but wants *brains*, *nostrils*, and *ears*,

The *eyes* are mostly two in number, without *eyelids*. They are either simple or compound, consisting of one or more lenses, and are the organs of vision in these as well as other animals.

Most insects have two *antennæ*, which are composed of an indefinite number of articulations; their use is as yet wholly unknown. They vary much in form, and are either

Setaceous,

Setaceous, growing gradually taper towards their point or extremity.

Filiform, resembling a thread, being throughout of equal thickness.

Moniliform, consisting of a series of knobs, like a necklace of beads.

Clavated, formed like a club, encreasing in thickness from the base to the point.

Capitated, encreasing in thickness towards their extremity, as the clubbed antennæ, from which they are distinguished by the form of their last or exterior articulation, which is larger and rounder than the others, forming a kind of *capitulum*, or *head*.

Fissile, which are like the last-mentioned, but have the *head* split or divided longitudinally into different *plates* or *laminæ*.

Pectinated, which have lateral appendices, resembling the teeth of a comb, or

Bearded, resembling a feather.

They are termed *short* (*breviores*) when shorter than the body, *midling* (*mediocres*) when of equal length with the body, and *long* (*longiores*) when longer than that part.

The *palpi*, by some called *feelers*, are articulated, fixed to the mouth, and generally either four or six in number, consisting of 2, 4, 3, joints: these seem to serve instead of hands to insects, they making use of them to approach their food to the mouth, and sustain it while eating.

The

The *mouth* is generally placed under the head, sometimes in the breast; it is furnished with a *rostrum* or *proboscis*, an *upper lip*, *jaws* placed transversally, *teeth*, a *tongue*, and a *palate*; some insects have no mouth.

The *stemmata* or *gems*, are three bright convex spots, or tubercules, placed upon the crown or upper part of the head.

The *trunk* is the part situate between the head and the abdomen; some of the feet are fixed to it; the upper part of it is called the *thorax*, behind which is the *scutellum*, or *escutcheon* (generally of a triangular form) for the insertion of which a piece appears to be cut out of the interior margin of each elytron: the under part is called the *sternum* and *breast*.

The *abdomen*, or *lower body*, contains the *stomach*, *intestines*, and *viscera*; it consists of five rings, or segments, and is pierced on the sides with *spiracula*, or *pores*, which supply the want of lungs; the upper part of it is called the *tergum*, or *back*, the under part the *venter*, or *belly*, which is terminated by the *anus*.

The *limbs* are the *tail* and the *feet*, to which (in many subjects) we may add the *wings*.

The *tail* terminates the abdomen; it sometimes has two appendices, or horns, and sometimes none; it is either *simple*, or *armed* with a *forceps*, a *fork*, a *bristle*, or a kind of *claw* or *sting*, which again is either *simple*, or *composed*, *smooth*, or *jagged* like a saw.

The *feet* are composed of *Femora*, or *thighs* (the joints immediately fixed to the body;) *tibiæ*, or *shanks* (the second joints) the *tarsi*, which form the third set, are composed of an indefinite number of articulations, and are terminated by the *ungues*, or *nails*: some have a kind of *hand* *(chela)*

(chela) or *claw*, with a moveable *thumb*; the hind feet are formed for executing different movements, as *running, leaping, swimming*.

The *wings* are, in some subjects, two, in others, four in number, and are either

Plain, stretched out their whole length without folds;

Plicatile, folded up;

Erect, sustained in an erect position, so that their extremities almost meet above the back of the insect;

Patent, open, expanded, extended, in an horizontal position;

Incumbent, covering horizontally the abdomen of the insect;

Deflected, in their position somewhat resembling the ridge of a house, declining downwards along the sides of the insect, but in such a manner

that

that the inner margins meet above the abdomen.

Reverſed, which differ from the laſt-mentioned, in the poſition of the under wings, theſe being placed horizontally, ſo that their edges project conſiderably from under the margin of the upper ones, which laſt are in the ſame direction as in the *deflected*.

Indented, with the edges cut out or ſcolloped.]

Caudated, in which one or more of the fibres of the wings are ſpread out or extended conſiderably beyond the margin, into a kind of tail.

Or *Reticulated*, when the veins or membranes of the wing croſs one another ſo as to reſemble net-work.

They are painted with *ſpots (maculæ)* bands *(faſciæ)* ſtreaks *(ſtrigæ)* which, when extended lengthways, are called *(lineæ)*

(*lineæ*) lines, and with *points* or *dots puncta*.

They are marked with *stigmates*, or spots, shaped like kidneys, and adorned with *ocelli*, or eyes, which consist of one or more rings (*the iris*) enclosing a spot *(the pupil)* which in general is of a different colour from the iris; these are either in their upper or under wings, and on the upper or under sides of the wings.

The *elytra*, or *wing-cases*, are two in number, of a crustaceous substance, and cover the under wings; they are for the most part moveable, and are either

Truncated, cut off at their extremity in a direct line.

Spinous, with spines or pointed elevations, Or

Serrated, with the exterior margin edged with spines, or teeth, like a saw.

Their superficies is either

Scabrous, rough.

Striated, marked with slight or shallow furrows.

Percated, with sharp longitudinal ridges.

Sulcated, deeply furrowed: Or

Punctuated, marked with concave or convex spots.

The upper wings, or *wing-cases,* are called *hemelytra,* when of a substance harder and stronger than the membranaceous wings which they cover, and yet softer than the *elytra* of the *Coleoptera.*

The *halteres (poisers)* are placed under the wings of *Dipterous insects,* or such as have but two wings, and probably serve to keep their bodies in equilibrio, when in the act of flying; they are composed of a head fixed at the end of a small pedicle or stalk.

As to sex, these animals are either *male* or *female*, which propagate their species; or *neuters*, which are incapable of generation, and seem to be devoted to the service of the other more perfect insects.

The *metamorphosis* in many insects, is threefold, and consists in a change of structure, effected by the subject casting the different coats in which the perfect insect is included, and as it were concealed.

The *egg*, containing the insect in its smallest size, or first state, is expelled from the ovary, as in other oviparous animals.

From the egg is produced the *larva*, or *caterpillar*, which is of a moist or humid substance, softer and larger than the egg, is without wings, sterile, or incapable of generation, slow in its motion, and is always exceedingly voracious when it meets with the food to which it is most addicted,
but

but more temperate when obliged to put up with that of which it is less fond. Many larvæ have a great number of *feet*, others have none.

The *pupa*, or *chrysalis*, is drier and harder than the larva, confined in a narrow compass, and is either *naked*, or *covered with a kind of web*; it often wants the mouth. Again, it is either

1. *Compleat*, having feet, and making use of all its limbs, *as the Spider (Aranea) the Tick (Acarus) the Woodlouse (Oniscus)*

2. *Semi-compleat*, or half compleat, which have feet, but only the rudiments, or, as it were, buds of wings, as the *Grasshopper (Gryllus)* the *Froghopper (Cicada)* the *Bug (Cimex)* the *Dragon-Fly (Libellula)* and the *Ephemera*.

3. *Incompleat*, having feet and wings, but which are immoveable, as in the *Bee*, the *Ant*, and the *Tipula*.

4. *Shrowded*,

4. *Shrowded (obtecta)* wrapped up in a cruftaceous covering, of fuch a form, that the part which contains the head and thorax may be diftinguifhed from that wherein the abdomen is lodged, as in Lepidopterous infects.

5. *Straitened (coarctata)* confined in a cafe of a globular make, not formed fo as to diftinguifh the different parts of the infect it contains, as in the *Mufca* (the *Fly)* and *Oeftrus* (the *Gad-fly.*

The infect, efcaped from its laft prifon, is in the third, or perfect ftate, is active, performs the work of generation, and is furnifhed with *antennæ,* which it generally wanted in its other forms.

The ftructure of the fame identical animal is therefore threefold, which fuppofes a like complication in the fcience, fince, in order to know it well, we muft be acquainted with the three different ftates through which it paffes.

Thefe

These animals are mute when not provided with some particular instrument separate and distinct from the mouth, with which they make a noise (as many do by the friction of some of their joints) and deaf, though they are by some means sensible of the vibration of the air; they are every where more in number than the species of existing plants, but seem fewer, on account of the greater field they have to range in. According to the climates they inhabit, they are either *tropical, arctical*, or *antarctical*, which last, however, are as yet unknown. In point of duration they are *annual* (except such as inhabit the waters) and, considered as individuals are the smallest of animals, but, taken all together, form the greatest part (with regard to bulk) of the animal kingdom. Their influence in the œconomy of nature is likewise the greatest, but being more generally diffused, and from their minuteness less obvious, is not so liable to be defeated, as if exercised by larger ani-animals; which security is the more necessary fary, as they are the yearly fervants of Nature, appointed in fufficient number for the perfecting fuch of her defigns as they are moft capable of accomplifhing, viz. preferving a due proportion among plants, confuming every thing that is mifplaced, fuperfluous, dead, or decayed in her productions; and, laftly, becoming nourifhment to other animals, and that chiefly to birds.

Infects are faid to *inhabit* thofe plants only upon which they feed, not thofe on which they fometimes may be met with, and trivial names, taken from that circumftance, are in general the beft, as being beft adapted to the purpofe of rendering *art* fubfervient to the explication of the views and police of *nature*. It is in confequence of thefe views and regulations, that we find fome infects occupied in preparing, others in purifying, others, again, in deftroying (according to the different apartments allotted them) the materials on which they work.

ORDINES INSECTORUM;

OR THE

ORDERS OF INSECTS.

INSECTS are divided into different orders, from the circumftance of their having or wanting wings, and from the number or fubftance of which thofe parts are compofed, in fuch as are furnifhed with them, as follows:

1. *Coleoptera.* Which have four wings; the upper ones called the Elytra, are entirely cruftaceous, being of a hard, horny fubftance, and join, or meet together, on the upper part of the body in a direct line or future.

2. *Hemiptera.* Which have four wings; the elytra differ from thofe of the former order in their hardnefs

2. *Hemiptera.* hardness, rather resembling strong parchment or vellum, than the horny substance of the Coleoptera; they cover the body horizontally; the inner margins extend the one over the other, not meeting in a direct line, as in the Coleoptera.

3. *Lepidoptera.* Which have four wings, all membranaceous, and imbricated, or covered with scales, fixed upon them nearly in the same manner as tiles are laid upon the roofs of houses.

4. *Neuroptera.* These have likewise four membranaceous wings, but which are naked, not being covered with scales as in the last mentioned genus; their abdomen is unarmed, or without a sting.

5. *Hymenoptera.* Which have four membranaceous naked wings, as the preceeding order, but the abdomen armed with a sting.

6. *Diptera.* Which have only two wings, being furnished with poisers or balancers, *Halteres*, instead of under wings.

7. *Aptera.* Or those which want wings.

The most distinguished writers who have formed Systems of Entomology besides our Author, are, (as I have observed in my Preface) Geoffroy, Scopoli, and Schæffer; each of these authors have pursued methods of arrangement very different from that of Linnæus, and from those of one another. I shall now proceed to give an account of their first, or general division, and shew wherein that differs from the orders invented and laid down as above, by Linnæus.

Geoffroy has divided this class of the animal kingdom into six sections only, uniting the Insecta Neuroptera and Hymenoptera of Linnæus, in his fourth, which accordingly consists of all such

such insects as have four naked membranaceous wings; these he has arranged under different articles or orders, according to the number of joints, or articulations, of which their feet are composed, rejecting Linnæus's division taken from the circumstance of their having or wanting stings, which, however, seems to argue them of very different natures and dispositions.

The order, or class, *Hymenopteron*, of Linnæus, indeed labours under one inconvenience, which may frequently mislead a beginner: I mean that of the male insects wanting the sting, or principal characteristic, which separates them from the Neuroptera. He will, however, soon learn to distinguish them from insects belonging to that genus, by the shape of their bodies, which, excepting those of some Ichneumons, are shorter, thicker, and stronger than the bodies of the Neuroptera; and particularly from the texture of their wings, in which the membranes run in general longitudinally, with very few cross ones: whereas the larger veins are so frequently crossed in the wings of the Neuroptera by small ones, as to make the wing resemble net-work.

The *Insecta Coleoptera*, or such as have the elytra of an horny or crustaceous substance, in their whole length, and the mouth armed with jaws, compose the first section of this author, which

which he has divided into three articles: The first, containing those insects whose elytra are crustaceous or horny, and cover the abdomen entirely: The second, those whose elytra are likewise crustaceous, but cover only a part of the abdomen: The third, those whose elytra are of a softer substance than the foregoing ones, and almost membranaceous; This last article comprehends such of the Linnæan *Insecta Hemiptera* as have the elytra, semi crustaceous in the whole length, or less hard, than those of the *Coleoptera*, and the mouth furnished with jaws, as the gryllus, or grasshopper, &c. This section is farther divided into orders, from the number of articulations found in the feet of the different insects which compose it.

His second section, or, *Insecta Hemiptera*, contains such of Linnæus's *Hemiptera*, as have elytra semi-crustaceous only to a certain distance from their base, as the Cimex or Bug, &c. but as this section, in which he has attempted to correct Linnæus, I think with success, would still, in that situation, have remained very incompleat, the Kermes and Coccus which he had referred to it, having only two wings, and those of the Psylla and Aphis being all four equally coriaceous; he has taken his essential character frm the proboscis or rostrum, with which the mouths of all the insects that compose it are furnished.

This

This proboscis, in most of the genera, is placed in the head of the insect, in others, (as the Psylla, the Kermes and Coccus) in the breast, between the first and second pair of legs.

In the third section, or *Insecta Lepidoptera*, he agrees entirely with Linnæus, as in his fifth, the *Insecta Diptera*, and his sixth, the *Insecta Aptera*.

In the Division of these sections into genera, he differs very much from Linnæus, as will be subsequently shewn.

Schæffer, who differs essentially from Linnæus, and in some things from Geoffroy, has divided his insects into classes, as follow:

1. Insecta Coleoptero-macroptera, or insects whose elytra are crustaceous in their whole length, and longer than the abdomen.

    This class comprehends the insects arranged by Geoffroy under the first article of the Coleoptera.

2. Insecta Coleoptero-microptera, differing from the former only in the length of their elytra, which, in this class, are not so long as the abdomen.

This clafs contains the infects which compofe the fecond article of the Coleoptera in Geoffroy.

3. Infecta Coleoptero-hymenoptera, feu Hemiptera, or fuch infects as have the elytra half cruftaceous, or becoming membranaceous towards their extremity.

4. Infecta Hymeno-lepidoptera, or with wings imbricated with fcales.

5. Infecta Hymeno-gymnoptera, or with naked and membranaceous wings; in this clafs he has not only followed Geoffroy in uniting the *Neuroptera* and *Hymenoptera* of Linnæus, but has moft unnaturally arranged the different kinds of Grylli, as grafshoppers, locufts, crickets, and the blattæ, or cockroaches, among wafps, bees, dragon-flies, and others of the fame nature.

6. Infecta Diptera, or infects having two wings; among thefe he has placed the *Coccus* and *Chermes*, which two genera feem to form a new clafs, differing from all others but the Diptera in the number of their wings, and from that genus in their want of halteres or ballancers.

7. In-

7. Infecta Aptera, or without wings.

The five firſt claſſes he has divided into orders, from the number of articulations in their feet; and the whole into genera, as will be hereafter noticed.

Scopoli agrees with Geoffroy in uniting all ſuch inſects as have the elytra cruſtaceous in their whole length, under the claſs *Coleopteron*. The *Grylli*, *Mantes*, and *Blattæ*, (*graſshoppers*, and *cockroaches*) ſeem however to form a claſs entirely diſtinct from the Coleoptera, from the different confiſtence or ſubſtance of their elytra; the ſhape of their heads, and the ſoftneſs of their bodies; and from the Hemiptera, in their having the mouth armed with jaws, nor extended into a proboſcis: Theſe reaſons may probably engage ſome future ſyſtematic writer to unite them in a new claſs, which may be termed *Inſecta Hemelytrato-maxilloſa*; preſerving to the *Hemiptera*, the name of *Inſecta Proboſcidea*, given to that claſs by Scopoli.

The other orders into which Scopoli has divided his inſects, are the ſame with thoſe of Linnæus; only to the fifth order or *Hymenoptera* of that author, he has given the name of *Inſecta Aculeata*, from their ſting; to the ſixth, or *Inſecta Diptera*, that of the *Halterata*, doubtleſs to diſtinguiſh

guish that order from the *Coccus* and *Chermes*, which have two wings, but want the halteres; and to the seventh, or *Aptera*, that of *Pedestria*.

ORDO I.

INSECTA COLEOPTERA.

This order is known by the cruftaceous elytra which cover the wings, and contains the following genera.

GENUS I. SCARABÆUS the BEETLE.

LINN. Syft. Nat. page 361.

The Scarabæus is diftinguifhed by the following characters. The ANTENNÆ, or horns, terminate in a kind of club, which is divided longitudinally into different plates, or laminæ, in fome feven, in moft three, in others two in number.

The fecond joint of the anterior or foremoft pair of legs, is armed with fpines or teeth.

Of this genus there are three sections or families, distinguished from one another as follows:

1. Those in which the thorax is armed with horns.

2. Those which have that part simple or unarmed, but which have horns on their heads.

3. Those in which the head and thorax are both simple or without horns.

Some of the insects belonging to each of these families, are *scutellati*, or furnished with the part called the *escutcheon*, and others belonging likewise to each of them, are *excutellati*, or want that part. This circumstance has induced Schæffer and Geoffroy to divide the Scarabæi into two genera, the one called *Scarabæus*, containing such as have the escutcheon, the other termed *Copris*, composed of those which want it.

The *Scarabæi* in each of these two last mentioned anthors, are divided into different families or sections, from the number of the plates or laminæ, of which the club that terminates the antennæ is composed.

The

The *Copres* are divided into families by Schæffer, in the same manner as Linnæus has divided his Scarabæi, viz. from their having or wanting horns on the head or thorax.

Scopoli has preserved the Linnæan genus entire, but has founded the divisions of it into sections, upon the number of spines, or teeth, with which the fore legs of the different species are armed.

The beetle called *the Bull comber*, and the two others mentioned beneath, are familiar instances of this genus.

The Larvæ, Caterpillars, or Grubs of many Scarabæi, lead a sedentary life under ground; most of these delight in, and feed upon dung, whilst others, particularly those from which the hairy Scarabæi, such as the *Garden Beetle* and *Cockchafer* are produced, live under, and consume the roots of plants; these last having compleated their metamorphosis, feed on the leaves of plants.

GENUS

GENUS II. LUCANUS the STAG-BEETLE.

LINN. Syst. Nat. page 559.

The antennæ of the Lucanus end or terminate in a club or knob, but of a different nature from that of the preceeding genus, the club being as it were compressed, or flattened on one side, which part thus compressed, is divided into short plates or laminæ, resembling the teeth of a comb.

The *Maxillæ*, or *Jaws*, are strong, porrected or advanced before the head, and are armed with teeth.

Schæffer and Geoffroy have given to this genus the name of *Platycerus*, without changing any of its characteristics. Geoffroy, however, has divided it into two families, from the form of the antennæ; the first family contains such as have the antennæ bent in the middle, and forming a kind of elbow or angle from the end of the first articulation, which, in this division, is as long as all the others: The second comprehends those whose antennæ are straight, or extended, with the first articulation of the same length as the others.

Scopoli

Scopoli agrees with Linnæus in name and characters.

The large Stag beetle is sufficiently known; its larva or grub, as most probably those of all the other *Lucani*, lives in rotten or decayed wood, and resembles those of the foregoing genus.

Genus III. Dermestes.

Linn. Syst. Nat. page 561.

The antennæ of the Dermestides end in a perfoliated club, or a head of an oval form, divided into different horizontal plates or leaves, which seem to be united together by a small stalk passing through their centre, and have three articulations thicker or larger sized than the others.

The thorax is of a convex form, and slightly margined.

The head is bent in, and as it were concealed under the thorax.

Schæffer and Geoffroy have taken from one of the Dermestides of our author the genus they have termed *Bostrichius*. This insect differs from the other Dermestides in the cubical shape of its thorax, Its antennæ are not perfoliated, but the three last articulations are much larger than the others.

The genus to which they have given the name of *Cistela* seems likewise to belong to the Linnæan Dermestis, from which it differs principally in

having

having six articulations of the antennæ larger than the others, and in the conical form of its thorax, which is likewise without any margin.

Geoffroy likewise adds to the characters by which Linnæus distinguishes this genus; that the last articulation of the antennæ is solid, which consideration, joined to that of the antennæ in several of the Linnæan silphæ, appearing rather to be perfoliated, than growing regularly thicker towards their extremity, probably induced him to refer such silphæ to this genus; these insects, however, differ much more essentially from the *Dermestides* than from the *Silphæ*, which last they perfectly resemble in their external appearance, in the flatness, breadth, and margin of their elytra, and the appendix or knob at the base of their hinder thighs, found upon all the Silphæ, and which Scopoli makes an essential characteristic of that genus.

Geoffroy has likewise placed some of the Linnæan Dermestides in which the last articulations of the antennæ are longer than in the others, among his *Byrrhi*, the *Linnæan Ptini*.

Scopoli has brought to this genus the Silpha Vespilio of Linnæus, on account of its antennæ, which are perfoliated. He observes, that this insect keeps the middle line between, or connects the two genera.

The larvæ, or maggots of the Dermestides, feed upon the carcases of dead animals, every kind of victuals, dried skins, the bark of trees, wood and seeds. Some of them make terrible havoc in collections of birds, insects, herbs, &c. These last resist the drugs generally made use of in museums for the destruction of insects, such as green wax, camphire, &c. but are killed by arsenic.

Genus IV. Ptinus.

Linn. Syst. Nat. page 565.

The antennæ of the Ptini are filiform: The last, or exterior articulations are longer than the others.

The thorax is nearly round, with a margin into which the head is received or drawn back.

Geoffroy has given the generical name of *Byrrhus* to some of the Ptini, in which he has observed the antennæ to be semi-clavated, or growing somewhat larger towards their extremity.

To the *Ptinus Pectinicornis*, Linn. No. 1. (which certainly differs much from the others of the same genus, in the form of the antennæ, they being (as its name infers) pectinated, and to another resembling it, he has given that of *Ptilinus*. That author likewise has placed the Ptinus *Fur*, Linn. No. 5. among his *Bruchi*, from the spherical form of its thorax.

Scopoli has placed the same insect among his Bupresides; he does not seem to have known the other insects belonging to this genus.

The

The larvæ or maggots of the *Ptini*, are found in the trunks of decayed trees, in old tables, chairs, &c. Some live and undergo their metamorphoses among hay, dried leaves, collections of dried plants, &c.

Genus V. Hister.

Linn. Syst. Nat. page 565.

The first articulation of the antennæ of this insect is compressed or flattened, and curve; the last, or terminating one, is considerably larger than the others, and appears to be a solid knob.

The head is drawn within the body, so that the jaws only appear.

The mouth is armed with jaws like a forceps.

The elytra are shorter than the body.

The fore legs are dentated, as in the Scarabæus.

Geoffroy and Schæffer have given the name of *Attelabus* to this genus, preserving all its characteristics, adding, however, that the antennæ are broken, or form an angle from the end of the first articulation, and that the feet are cursorii, or made for running.

The first has observed that the capitulum, or knob of the antennæ which appears to be solid, is composed of several rings or circles strongly united together, but which the insect can separate and display, or contract at pleasure.

Scopoli agrees with Linnæus likewise in name.

The insects belonging to this genus, as well as their larvæ are frequently met with in the dung of horses, cows, &c.

Genus VI. Gyrinus.

Linn. Syst. Nat. page 567.

The antennæ of this insect are club-formed, stiff, and shorter than the head.

It has four eyes, two on the upper, and two on the under side of the head.

Geoffroy adds to the above characters that the feet are natatorii or formed for swimming.

Scopoli has arranged the Gyrinus along with the Dytisci, from which it differs essentially in the number of its eyes and the form of its antennæ; these indeed in some of the Dytisci are clubbed, but the club is perfoliated, nor are the antennæ stiff as in the Gyrinus.

The insect called the *Water-flea* belonging to this genus, is very frequently met with in standing waters, and easily distinguished by its shining black colour, and the swiftness and circular direction of its motion in swimming.

I do not know that its larva has yet been observed, but it may probably be found along with that of the Dytiscus, which without doubt it resembles.

Genus VII. Byrrhus.

Linn. Syst. Nat. page 568.

The antennæ of the Byrrhus are club-formed, and terminate in a capitulum or knob, which is of an oval form, rather compressed or flattened, and almost of a solid substance, (sub solidum.)

Geoffroy and Schæffer agree with Linnæus in the definition of this genus, to which they have given the name of *Anthrenus*, the insects belonging to it being generally found upon flowers.

Schæffer has added to the characters assigned to it by our author, that the head is bent, or inclined downwards, and hid under the thorax; which particularity is of great service in distinguishing this insect, the form of the antennæ alone being scarce sufficient for that purpose.

Geoffroy observes that the larvæ of the Anthreni are found upon plants, or in the bodies of half decayed animals; they often undergo their metamorphosis in the bodies of preserved insects, which they reduce to powder.

Genus VIII. Silpha.

Linn. Syst. Nat. page 569.

The antennæ of the Silphæ are small at their base, and grow insensibly thicker towards the end.

The elytra have a margin.

The head is prominent.

The thorax is rather flattened, with a margin.

Schæffer has composed two genera from the Silphæ of our author. The one named *Silpha*, containing such Linnæan Silphæ as have the margins of the head and thorax most apparent, and the thorax more convex: The other, called *Peltis*, composed of those in which the margin of the elytra is less apparent, and the thorax flatter than in the others.

Geoffroy has arranged several of those insects among his Dermestides, and of the others has formed the genus Peltis, containing such as have the thorax and elytra more strongly margined, and whose antennæ appear to be be perfoliated.

Scopoli

Scopoli adds to the Linnæan characters of the Silphæ their having a kind of lamina or knob, which terminates in a spine, situate at the base of their hinder thighs,

Many of the Silphæ are found early in the spring, under the loose bark of trees, and they, as well as their larvæ, feed chiefly on the half-decayed carcases of animals.

Genus IX. Cassida, the Tortoise Beetle.

Linn. Syſt. Nat. page 574.

The antennæ of the Caſſida are nearly filiform, but grow ſomewhat thicker towards the end.

The elytra have a broad margin.

The head is entirely concealed under the thorax.

The thorax is flat and margined, forming a kind of ſhield for the head.

The inſect called the Green Tortoiſe Beetle, belongs to this genus.

The larvæ of the Caſſida eat the under ſide of the leaves of plants, and often, as it were, hide themſelves under a cover of their own excrements, ſupported in the air above their bodies, by means of their forked tail.

Schæffer and Geoffroy have adopted this genus without any alterations. The latter obſerves, that the antennæ are *nodoſæ*, knotty, or compoſed of large articulations. Scopoli has referred to it the Lampyrides *Noctiluca* and *Sanguinea*,

nea, though these two insects seem to differ much from the Cassidæ in the form of the segments of their belly, which terminate on each side in round and soft appendices; the belly of the Cassida on the contrary is simple.

The oblong form and flatness of the abdomen in the lampyrides serves likewise to distinguish them from the Cassida, which last are almost oval, with the abdomen much more elevated in the middle than on the sides; from which circumstance the name of Tortoise Beetle has been given to it in our language.

GENUS

Genus X. Coccinella.

Linn. Syft. Nat. page 579.

The antennæ of the *Coccinella* are fubclavated, or increafe a little in thicknefs towards the end. The laft joint appears as if the end of it was chopped off.

The palpi are club-formed, the laft articulation being fhaped fomewhat like a heart.

The body is hemifpherical.

The thorax and elytra are margined.

The abdomen, or belly, is flat.

This genus is fubdivided into fections from the colour of the elytra, and of the fpots with which they are adorned, as follows:

1. Thofe whofe elytra are red or yellow, with black fpots.

2. Thofe fpotted with white, on a red or yellow ground.

3. Thofe

3. Those with black elytra spotted with red.

4. Those with black elytra, and white or yellow spots.

Scopoli says that the *Coccinella* differs chiefly from the *Chryfomela* in the length of the antennæ, those of the Coccinella being shorter than the thorax, but in the Chryfomela twice the length of that part. The antennæ differ likewise in shape, those of the last mentioned genus being filiform, or throughout of equal thickness, whereas those of the Coccinella grow thicker towards the end.

Schæffer and Geoffroy agree with Linnæus in the characters of this genus.

The larvæ of the Coccinellæ devour the Aphides, and by that means contribute to cure plants which those animals infest, of the *Phithiriafis*, or loufy disease.

Genus XI. Chrysomela.

Linn. Syſt. Nat. page 586.

The antennæ of the Chryſomela are compoſed of little globular articulations which grow larger towards the end; and somewhat reſemble a necklace of beads.

Neither the thorax nor elytra have any margin.

Linnæus has divided this genus into families, as follows:

1. Thoſe whoſe bodies are of an oval form.

2. Thoſe whoſe hinder thighs are much thicker than the others, being ſaltatoriæ, or made for leaping.

3. Thoſe whoſe bodies are of a cylindrical form.

4. Thoſe of an oblong form, and in which the thorax is broader or wider than the abdomen.

5. Thoſe

5 Those which are long, of a slender make, and which have the thorax of equal breadth with the abdomen.

Linnæus observes that this last mentioned family differs a little from the preceding ones, being more oblong, and the body more elevated in the middle than on the sides, but that he had not been able to discover the limits by which they should be distinguished, nor any other genus under which they could be more properly arranged.

From these different kinds of Chrysomelæ GEOFFROY has formed several genera, viz.

The *Galeruca*, which differs from the other Linnæan Chrysomelæ in the roughness and margin of its thorax.

The *Chrysomela*, whose thorax is smooth and margined.

The *Cryptocephalus*, the articulations of whose antennæ are rather longer than in those of the other Linnæan Chrysomelæ, and the thorax of an hemispherical form.

The *Crioceris*, which differs from the other genera in the cylindrical form of its thorax.

The *Diaperis*, the articulations of whose antennæ being rather larger than in the other species of the Linnæan Chryfomelæ, appear to be perfoliated. The thorax in this genus, of which he has only one species, is convex and margined.

The *Altica*, which genus comprehends that family of the Linnæan Chryfomelæ, whose hinder thighs are made for leaping.

The *Melolontha*, whose antennæ are ferrated, or with lateral appendices like a faw, and placed on the fore part of the head before the eyes.

SCHÆFFER has followed Geoffroy in thefe alterations, adding, that the head of the *Criptocephalus* is drawn back within the thorax; that of the *Crioeeris*, on the contrary, is ftretched forwards, or porrected.

SCOPOLI has arranged fuch of the Linnæan *oval* Chryfomelæ as have the antennæ fcarce fo long as the thorax, among his Coccinellæ, others, whose heads appear to be a little drawn in, or, as it were, half hid under the thorax, among his bupeftrides; and thofe of the fourth divifion, in which the thorax is rather broader than the head and body, among his Attelabi.

The

The distinctions from which Geoffroy and Schæffer have formed so many new genera, are in general too trifling to be taken for *generical*; in which case, the multiplication of genera, instead of elucidating the science, serves but to render it more obscure.

The larvæ of the Chrysomelæ consume the pulp of leaves, rejecting the fibres: Those of the Chrysomela Saltatoriæ infest the cotyledons and tender leaves of plants.

The insect called the Lady Cow, or Lady Bird, belongs to this genus.

Genus XII. Hispa.

Linn. Syst. Nat. page 609.

The antennæ in this genus are fusiform, growing gradually larger from each extremity towards the middle: they are situate between the eyes, and are placed so near one another at their base, as to seem to arise from the same point.

The thorax and elytra are in general covered with protuberances or spines.

Geoffroy has placed the only species belonging to this genus, which he had met with in France, among his *Crioceres*, the oblong Chrysomelæ of Linnæus. The shape of the antennæ and their situation, however, sufficiently distinguish the Hispa from that genus.

The larva of the Hispa seems to be yet wholly unknown; there are but two species of the perfect insect found in Europe, and they are to be met with at the roots, or on the blades of different kinds of grass.

Genus XIII. Bruchus.

Linn. Syst. Nat. page 604.

The antennæ of the Bruchi are filiform, growing thicker towards their extremity.

Linnæus's definition of this genus is comprehended in these few words; and the two circumstances from which the insect is to be discovered contradictory, as filiform antennæ are throughout of equal size; neither does he give such a description of any one of the species arranged under it as can enable us to distinguish the Bruchus from other genera. I have only seen one species, the Bruchus Pisi, in which the antennæ are placed exactly before the eyes, and are composed of triangular articulations growing larger towards their extremity, with the last one of an oval form. It has four palpi seated at the extremity of a proboscis which is rather broader than it is long. The elytra are rounded at their extremity, and a fourth part shorter than the abdomen. Whether or no these are generical characters, by which the other insects belonging to the *Bruchus* may be distinguished, will best be observed by those who possess a greater number of the species described by Linnæus. This insect is arranged by Scopoli under the genus termed ed by him *Laria*, to which he assigns the following characters: The antennæ larger towards their extremity; the thorax elevated in the middle and rounded towards the sides; the knob situate at the base of the thighs in the Silpha is wanting in this genus.

The same insect is placed by Geoffroy with his Mylabres, which genus, he says, equally resembles his Chrysomelæ, and the Linnæan Curculiones, connecting the two genera.

Genus XIV. Curculio.

Linn. Syſt. Nat. page 506.

The antennæ of the Curculio are ſub-clavated, and ſeated in a roſtrum or probofcis, which is of a horny ſubſtance, and prominent.

The Curculiones are divided into the following ſections:

1. Thoſe which have the roſtrum longer than the thorax, and whoſe thighs are ſimple, without teeth or ſpines.

2. Thoſe in which the roſtrum is longer than the thorax, and the thighs dentated.

3. Thoſe which have dentated thighs, and the roſtrum ſhorter than the thorax.

4. Thoſe whoſe thighs are ſimple, and roſtrum ſhorter thau the thorax.

Scopoli obſerves that the Curculio is a ſluggiſh inſect, and that it endeavours to eſcape its foes by contracting its members and letting itſelf fall to the ground. That author (who diſtinguiſhes this genus by the same characters as Linnæus) has divided it into two families, the first whereof comprehends those which have straight or extended antennæ; this family is sub-divided into the following sections:

1. Those in which the rostrum is thicker than the thighs and shorter than the thorax; among these he has placed some Linnæan *Attelabi*.

2. Those which have the rostrum thicker than the thighs and longer than the thorax.

3. Those in which the rostrum is smaller than the thighs and longer than the thorax; the thighs in some of the insects belonging to this section are *dentated*, in others, *muticæ*, or without spines.

The second family consists of those whose antennæ are bent or form an angle, and contains the following sub-divisions:

1. Those with the rostrum larger than the thighs, which are *spinosæ*, or armed with spines.

2. Those with the rostrum as in the other, but without spines on the thighs.

3. Those with the rostrum smaller than the thighs, which are unarmed, or without spines.

Geoffroy

Geoffroy divides this genus (which with him is limited to such Linnæan Curculiones as have antennæ bent, or forming an angle in their middle) into two families, from the circumstance of the thighs being armed with, or wanting spines.

To others of them which have extended, or straight antennæ, (those belonging to the first family of Scopoli's Curculiones) he has given the generical name of *Rhinomacer*, under which genus he has likewise arranged some Linnæan Attelabi.

The genus named by him *Mylabris*, seems to belong to the Curculio of our author; he has distinguished it by the following characters.

The antennæ growing larger towards the end composed of hemispherical articulations, and placed upon a short and broad rostrum or proboscis.

Four small antennæ (perhaps palpi) placed at the extremity of the proboscis.

Schæffer has followed Geoffroy in these divisions of the Linnæan Curculiones.

The larvæ of the long beaked Curculiones live upon fruits, feeds of different plants, and corn, often making terrible havoc in granaries.

Thofe of the fhort beaked ones devour the leaves of plants; many of them pierce and lodge in the ftalks.

The infect called the *Weevil* by farmers, belongs to this genus.

Genus XV. Attelabus.

Linn. Syst. Nat. page 619.

The Attelabus is distinguished by the shape of its head, which is broader in the fore part (occasioned by the prominency of the eyes) than behind, or which tapers gradually from the eyes towards the thorax.

The antennæ are thicker towards their extremity than at their base.

This genus Linnæus observes is very obscure, the insects arranged under it differing much from one another in their external appearance. This obscurity I imagine however rather to proceed from his not having known a sufficient number of insects proper to be arranged under it, and his placing with those that are, some others, (as the *Clerus* of Geoffroy) in which the generical characters he assigns to it are not found, rather than from any defect in the characters themselves, having lately observed in different collections many exotic insects which answer most exactly his definition of the Attelabus. If some insects which he has referred to it, were rejected, the genus, I think, would be very distinguishable, and sufficiently numerous.

Scopoli

Scopoli distinguishes the Attelabi by the following characters.

The hinder part of the head gradually diminishing in size.

The eyes prominent.

The thorax somewhat broader than the diameter of the head, taken from one eye to the other, and of a more cylindrical form.

Among these he has arranged some of the Linnæan Chrysomelæ, whose bodies are oblong, and narrower than the thorax.

Some of the Linnæan Attelabi are placed by him among his Curculiones.

The *Clerus* of Geoffroy and Schæffer is taken partly from this genus, and partly from the Dermestes of our author. They have given to that genus the following characters:

The antennæ club-formed, and placed on the head, the knob composed of three articulations.

No proboscis.

The thorax almost cylindrical, without any margin.

The under fide, or plant of the feet, fpongy.

They have arranged fuch Attelabi as moft refemble the Curculiones, under the genus *Rhinomacer*, from which, however, thefe feem to differ effentially in the fituation of the antennæ, which, in the *Attelabus* are placed upon the head, but in the *Rhinomacer* upon the roftrum.

The larvæ of many of the Attelabi refemble fo much thofe of the Curculiones as not to be diftinguifhed from them without difficulty.

Genus XVI. Cerambyx.

Linn. Syst. Nat. Pag. 621.

The antennæ of the Cerambyces are composed of articulations, which gradually diminish in size as they approach towards, or are situate nearer to the extremity.

The Thorax is either armed with spines or gibbous made uneven by small elevations.

The Elytra are narrow, and throughout of equal breadth.

This Genus is divided into sections, from the form of the thorax, and that part being or not being armed with spines, as follows:

1. Those which have the thorax armed on each side with moveable spines.

2. Those in which the thorax is margined, and sides armed with spines.

3. Those in which the thorax is round, and armed with fixed spines.

4. Those which have the thorax nearly of a cylindrical form, and unarmed, or without spines.

5. Those which have the thorax of a roundish form, resembling a globe flattened or depressed on the upper side.

Scopoli has assigned the power of emitting a sound or noise, by the friction of the thorax, where joined to the body, as a character of the Cerambyx; this vague definition has occasioned his placing several of the Linnæan Cerambyces, which want that property, among his Lepturæ: he makes only two divisions of the remaining Cerambyces, the first containing such as have the thorax armed with spines; the other, those in which that part is unarmed; this method is more simple than that of Linnæus, and perhaps as proper, in collections consisting wholly of European insects.

Geoffroy and Schæffer have formed several new genera from the different kinds of Cerambyces.

To those which have serrated antennæ placed in the eyes, or surrounded at their base by the eyes, they have given the generic name *Prionus*.

To those whose antennæ grow gradually taper, from the base towards their extremity, and are placed in the eye, they have preserved the name of *Cerambyx*; the thorax in this genus is armed with spines.

Others with setaceous antennæ placed in the eyes, and the thorax of a cylindrical form, without spines, they have arranged along with their Lepturæ.

The antennæ in their *stenocorus* taper towards their extremity, as those of the Cerambyx, but they are placed before the eyes, and the elytra diminish in breadth towards their point. This genus is divided into two families, the first of which only belong to the Linnæan Cerambyces, being such as have the thorax armed with spines, the other, in which the thorax is unarmed, belongs to the *Leptura* of our author.

The insect generally known by the name of the Goat-Chafer, or Musk-Beetle, is a Cerambyx, and as its thorax is round, and

found

armed with fixed spines, it must belong to the third family of our author. It is frequently found on the willow in the autumn, and smells like musk, from which circumstance its name is taken.

The larvæ of the Cerambyces nourish themselves with the interior substance of trees, into which they penetrate, and where they live and perform their metamorphosis.

Genus XVII. Leptura.

Linn. Syst. Nat. Pag. 637.

The antennæ of the Lepturæ are setaceous, growing gradually taper towards the end.

The Elytra diminish in breadth towards their extremiy.

The thorax is of a roundish and slender make.

This genus is divided into two sections, the first containing those in which the thorax is somewhat oblong, but broader at its base than where joined to the head, and whose elytra are truncated or cut off at their extremity, in a direct line; the second comprehends those in which the thorax is nearly of a globular form, and whose elytra are obtuse at their extremities.

Scopoli observes, that the elytra of the Lepturæ are stiff, nor flexible as in the Cantharis.

The Genus, termed by Geoffroy *Leptura*, is composed of such Linnæan Cerambyces as have setaceous antennæ, surrounded at their base by the eyes, and the thorax naked or without spines, and such of the Lepturæ of our author as have their antennæ situate in the eye: in this he is followed by Schæffer; the remaining Lepturæ are referred by these two authors to their *stenocorus*, as before observed.

The larvæ of this genus are found with those of the preceding one, and much resemble them in outward appearance and way of life.

Dr. Berkenhout has called some of the Linnæan Lepturæ *Wasp Beetles*. I am not certain whether they are generally known by that name.

Genus XVIII. Necydalis.

Linn. Syst. Nat. page 640.

The antennæ of the Necydalis are setaceous, as in the foregoing genus.

The elytra are either shorter than the abdomen, or narrower, and of the same length with that part.

This genus is divided into two families: The first containing those which have elytra shorter than the wings and abdomen; the other those in which the elytra are as long as the body, but narrower, being shaped like an awl, or drawn to a point, and a little curve at their extremities.

Schæffer has confined the genus *Necydalis*, to one insect, the Necydalis Major, Linn. No. 1. The others belonging to the same section, he has arranged under his *Mylabris*, on account of their antennæ, which according to him are filiform, and placed upon a short proboscis; the Necydalis of the second family or section, he has arranged among his Lepturæ, from their antennæ being seated in the eyes.

These last are placed by Scopoli among his Cantharides.

The insects belonging to the first division of this genus, differ from the Staphilini in the want of the little vesicles, or bladders, which these last frequently thrust, or shoot out of the hinder part of their abdomen, when in distress, and in their antennæ; they differ from all the other Coleopterous insects, in their wings being extended their whole length, nor folded up under the elytra, which, on that account, seem to be of less use to the *Necydalis* than to the other genera belonging to that order, since only so much of the wing as is covered by the elytron can be preserved by it.

I do not find that the larva of the Necydalis has been known to any author.

Genus XIX. Lampyris.

Linn. Syst. Nat. page 643.

The antennæ of the Lampyris are filiform.

The elytra are weak and flexible.

The thorax is flat, and of a semiorbicular form, surrounding and concealing the head.

The segments of the abdomen terminate on each side in *papillæ*, or little appendices, which turn, or are bent upwards towards the elytra, and in part cover one another.

The females, in general, want wings.

Scopoli, who has only described two species of this genus, has arranged them with the Cassida of Linnæus, giving to that genus the simple characteristic of the head being concealed under the thorax. That character the Lampyris has in common with the Cassida, from which, however it differs in the length and flatness of the body, in the shape of the antennæ, which in the Cassida grow thicker towards their extremity, and in the papillæ, or folds of the abdomen, which

are

are wanting in the laſt mentioned genus, and ſerve more particularly to diſtinguiſh the Lampyris.

Geoffroy and Schæffer give the ſame characters to this genus as Linnæus.

The *Pyrochora* of the laſt mentioned author is a Linnæan Lampyris, with antennæ pectinated on the one ſide.

The larvæ of thoſe Lampyrides we are acquainted with, perfectly reſemble the female inſect, and feed upon leaves.

The inſect called in our language, the glow-worm, from the ſhining light which it emits, and which is ſo frequently met with in the evenings about the month of June, in woods and meadows, belongs to this genus.

Genus XX. Cantharis.

Linn. Syſt. Nat. page 647.

The antennæ of the Cantharis are ſetaceous.

The thorax is margined, and ſhorter than the head.

The elytra are flexible.

The ſides of the abdomen are edged with papillæ, or appendices, folded upwards, as in the preceding genus.

The Cantharides are divided into two ſections; the firſt diſtinguiſhed by the flatneſs and breadth of the thorax, which part in the other is rounded on the ſides and narrower.

Scopoli, who deſcribes under the ſame generic title ſuch of the Linnæan Cantharides as he had found in his country, and which all belong to the firſt ſection of our author, obſerves, that the thorax, under which a part of the head is concealed, is of a convex form.

Geoffroy

Geoffroy has given the generical name of *Cicindela* to such of the Linnæan Cantharides as he has described. He differs from Linnæus in his opinion of the form of the antennæ; which, according to him, are filiform rather than setaceous.

His *Pyrochroa* is a Linnæan Cantharis, with pectinated antennæ. The generical name of *Cantharis* he has given to the winged *Meloes* of Linnæus, or those of his second section.

Schæffer has given the generical name of *Telephorus* to some Linnæan Cantharides, which differ from the others in the number of the articulations of which their tarsi are composed. He has placed others of them, in which the antennæ are seated in the eyes, and the thorax flat, with a less perceptible margin than in the others, among his Lepturæ.

The larva of the Cantharis was almost unknown to Linnæus, and wholly so to Geoffroy. My ingenious friend Mr. Curtis has lately discovered it, and observed the metamorphosis of some of them; they resemble those of the Cerambyx, and were found in the decayed trunk of a willow.

Genus XXI. Elater.

Linn. Syst. Nat. page 651.

The antennæ of the Elater are setaceous.

An elastic spring or spine projects from the hinder extremity of the breast or under side of the thorax.

The insect, when laid upon its back, rises and sustains itself upon the anterior part of its head, and the end, or point, of its abdomen or elytra, by which means the spine of its breast is withdrawn from out of a groove or cavity of the abdomen, where it is lodged when in its natural position; then suddenly bending its body, the spine is struck with force across a small ridge, or elevation, into the cavity from whence it was withdrawn, by which shock, the parts of the body before sustained in the air, are so forcibly beat against whatever the insect laid upon, as to cause it to spring, or rebound, to a considerable distance.

Geoffroy

Geoffroy obferves, that a cavity is fcooped out of the under fide of the head and thorax of the Elater, in which the antennæ are lodged, probably to preferve them from the violence of the fall, when it makes the fingular leap which diftinguifhes it from all other infects.

The character taken from the antennæ by our author is extremely vague, for, as Schæffer juftly obferves, they are in fome fetaceous, in others filiform; fometimes they are pectinated, and fometimes ferrated; the fpines at the extremity of the thorax are, however, fufficient marks to diftinguifh them by, being found upon almoft every one of them, and rarely met with in any other of the Coleopterous order of infects. Scopoli has called one of his Elateres *Degener*, becaufe it differs from the others, in the want of thofe fpines, the hinder part of its thorax being round. Such are beft diftinguifhed from the Bupreftis (which genus the Elater moft refembles) by the elaftic fpine, fituate at the extremity of the breaft.

Schæffer likewife obferves, that the hinder angles of the thorax are very much pointed or extended into fpines, and that the tarfi have five articulations, or joints.

Linnæus

Linnæus was unacquainted with the larva of the Elater, but we learn from Geoffroy, that it lives and undergoes its metamorphosis in the trunks of decayed trees.

That author, however, has said nothing with regard to its formation, so that we are still ignorant whether or no it resembles that of the Buprestis. The compleat insects are frequently found on flowers and plants; some of them frequent the banks of running waters, sandy banks, &c. and are pretty well known. They are in some places not improperly called *Skippers*.

Genus XXII. Cicindela.

Linn. Syst. Nat. page 657.

The antennæ of the Cicindelæ are setaceous.

The maxillæ, or jaws, advance considerably before the head, and are armed with teeth.

The eyes are rather prominent.

The thorax is roundish and margined.

Scopoli and Schæffer observe that the Cicindela have an obtuse lamina, or knob, at the base of the hindermost thighs, and that the head is broader than the thorax; which circumstance is chiefly occasioned by the prominency of the eyes.

Geoffroy has arranged such insects belonging to this genus of our author as he has described among his *Buprestides,* (the Linnæan *Carabi*) from which the Cicindela principally differs in the form of the thorax, which in it, is roundish, but in the Carabus of the form of a heart, and cut off at the end in a direct line. This difference,

ence, however, he reconciles, by dividing his Bupreftides into two families, diftinguifhed from one another by thefe circumftances. The eyes of the Cicindela are much more prominent than thofe of the Carabi.

Schæffer adds to the characters given by our author to this genus as above, that the jaws are crooked, and the feet made for running.

The larvæ of this genus live chiefly with thofe of the Carabi, in deep holes under the earth, and as well as the perfect infects, devour weaker animals for their food.

Genus XXIII. Buprestis.

Linn. Syft. Nat. page 659.

The antennæ of the Bupreftis are fetaceous, and as long as the thorax.

The head is half retracted, or drawn back within the thorax.

They are divided into three families, diftinguifhed by the following marks.

The elytra in the firft decline towards the fides, being much elevated at the future, and particularly fo, near their bafe.

In the fecond, they are ferrated, or armed with fhort fpines, near their extremity.

In the third, they are whole, or entire.

Scopoli has arranged fuch of the Bupreftides of our author as he knew, among his Mordellæ, of which genus he fays nothing more than that they have an appendix or broad plate, which covers and defends the hindmoft thighs, forming a kind of cavity into which they are received.

The

The genus to which that author has given the name of *Bupreſtis*, confiſts chiefly of the oblong Chryſomelæ of Linnæus as before obſerved. He diſtinguiſhes that genus by the following characters.

The antennæ never ſhorter than the thorax.

The head deflected, half drawn back within the thorax.

The thorax as it were ſwelled, or puffed up like a cuſhion, *(pulvinatus)*.

The abdomen obtuſe.

According to him the other Linnean Chryſomelæ differ from thoſe arranged with his Bupreſtides in their heads, (which are porrected or advanced before the thorax) being leſs thick or bulky; and in the antennæ, which in the Chryſomelæ are twice the length of the thorax.

The antennæ are generally ſerrated in this genus, as obſerved by Geoffroy, who has given the generical name of *Cucujus* to the French Bupreſtides.

Schæffer ſays that the mouth of the Bupreſtis is armed with jaws and palpi, that the tarſi have five articulations, and that the elytra are margined, and cover the abdomen.

The *Bupreſtis* and *Elater* resemble one another very much, and are best distinguished by the spines, which terminate the breast and thorax of the latter.

There are but few species of this genus found in Europe, and we are wholly unacquainted with their larvæ and metamorphosis; they are generally of bright shining colours, from which circumstance Geoffroy has chosen the generical name which he has given them.

Genus XXIV. Dytiscus.

Linn. Syst. Nat. page 665.

The antennæ of the Dytiscus are either setaceous, or increase in size towards the end, and have a perfoliated capitulum or head.

The hind feet are hairy, made for swimming, and are armed with small claws.

This genus is divided into two families; the first composed of those which have perfoliated antennæ; the second of those in which the antennæ are setaceous.

Geoffroy has formed from the two Linnæan families of Dytisci, as many genera. To those with perfoliated antennæ (which he says are shorter than the palpi) he gives the generical name of *Hydrophilus*; the others, in which, according to him the antennæ are filiform, and longer than the head, he calls *Dytici*.

Schæffer, who has adopted this division of the Linnean genus, says, that the tarsi of the Dytiscus have five articulations; that the body is
oblong

oblong, and the head obtuse; the mouth of the Hydrophilus, according to the same author, is armed with jaws, and has four palpi, two of which are longer, and two shorter than the antennæ.

Scopoli observes that the Dytiscus is a dull and sluggish insect.

The plants, or under side of the fore feet of the male Dytisci are hemispherical. The elytra of the females are generally furrowed. The first resemble the Dermestides; the females are more like the Carabi: It is very difficult to distinguish the sex or species. Their larvæ are frequently met with in ditches, they are not to be bred, or do not go through their metamorphosis when confined, without great difficulty; and if two or three are kept together in a small place, never fail to devour one another. Many species of the compleat insect are very common in stagnated waters, which they quit in the evening to fly about. They are known by the name of *Water Beetles.*

Genus XXV. Carabus.

Linn. Syst. Nat. page 631.

The antennæ of the Carabus are setaceous.

The thorax is shaped somewhat like a a heart, the point of which is cut off, and is margined.

The elytra have a margin.

The Carabi differ greatly in size, and are divided from that circumstance, into two families; the first containing the larger, the second the smaller ones.

Geoffroy has united the Cicindela of our author with this genus, under the generical name of *Buprestis*, given by Linnæus to another genus; he adds to the above characters of the Carabus, that they have a considerable lamina or knob at the base of the thighs, which is found also in the Cicindela, but is scarce sufficient to justify the placing that insect under the same genus with the Carabus, from which it differs in the prominency of the eyes, and the roundness of its thorax: The same knob is found at the base of the thighs of the Silpha, and some other insects.

The same author asserts, that the antennæ in this genus, and likewise in the Cicindela, are filiform rather than setaceous, which is sometimes observable in European subjects, but generally in those as well as in exotic ones, they taper towards the point.

Schæffer observes, that the head of the Carabus is prominent, the mouth armed with jaws, and four articulated palpi, and that the feet are made for running. The tarsi in all the feet are composed of five articulations.

Scopoli, who divides this genus into families from the same circumstance as Linnæus, fixes the length of the first or greater ones at seven lines.

The larvæ of the Carabi live in the ground or in decayed wood, where they perform their metamorphosis; they themselves live chiefly on weaker insects, or small larvæ.

The name of *Ground Beetle* has been given by some authors to the Carabus; others have called it the *Blaine Worm*.

Genus XXVI. Tenebrio.

Linn. Syst. Nat. Pag. 674.

The antennæ of the Tenebrio are moniliform, or resemble a string of beads: the ultimate articulation is rounder than the others.

The thorax has a margin, and is of a convex form, though rather flattish, the elevation being inconsiderable.

The head is porrected, or stretched forwards.

The Elytra are rather stiff.

This genus is divided into two sections; the first containing such Tenebriones as want wings, and in which the elytra are united, or without a longitudinal suture; the second, such as are furnished with wings.

According

According to Scopoli, the antennæ in this genus are always longer than the thorax: he alſo obſerves, that many of the Tenebriones very much reſemble the Carabi, but are diſtinguiſhed by the antennæ, and by the lamina at the baſe of the thighs, in the Carabi, which is never found in the Tenebrio; to which add, that the abdomen of this laſt is more oblong, and not ſo flat as that of the Carabus; and that the tarſi of its hind feet have only four articulations.

Geoffroy obſerves, that the antennæ in ſome of the Tenebriones are compoſed of long articulations, which are throughout of equal ſize, in others, of globular, or oblong ones, growing larger towards their extremity, and from this circumſtance he has divided them into two families, in which he is followed by Schæffer.

Scopoli has preferred the method of Linnæus.

The larvæ of the Tenebriones are frequently met with under heaps of weeds, branches of trees and other refuſe of gardens; ſome of them live under ground, others in meal, neglected and dry bread, &c. The compleat inſects are found in houſes, gardens, and ſandy places; they run with great ſwiftneſs, and generally emit a very fœtid ſmell; they are, on that account, ſometimes called *ſtinking Beetles*. One ſpecies, frequently found in houſes, is called the *ſlow-legged Beetle*.

Genus XXVII. Meloe.

Linn. Syst. Nat. Pag. 679.

The antennæ of the Meloe, like those of the preceding genus, resemble a string of beads, but the last articulation, which in the Tenebrio is round, in this genus is of an oblong oval form.

The thorax is roundish.

The elytra are soft and flexible.

The head is inflected and gibbous.

Many of the Meloes want wings, with which others of them are furnished; they are divided into two families; the first containing those which are apterous, and have elytra shorter than the abdomen; the second, those which are winged and have elytra as long as the body, by which the wings are wholly covered.

Scopoli adds to the Linnæan characters of the Meloe, that the thorax tapers, or grows slenderer from its middle towards each each extremity.

Linnæus has united with this genus the *Notoxus* of Geoffroy, remarkable from the horn upon its thorax; *Vid: Linn. Mel. No.* 14. Geoffroy asserts that its antennæ are filiform, which circumstance should seem to separate it from the Meloe, to which, however, our author, who appears *very* unwilling to multiply the genera of insects from trivial circumstances, thinks it resembles more than to any other.

Geoffroy has separated the *Meloe Schæfferi Linn. No.* 12, from the other species of our author, on account of its antennæ, and has given to it the generical name of *Cerocoma*: according to him the antennæ of the female are composed of eleven articulations, the ten first of which are very short, and the eleventh, or exterior one, at least as long as a third part of the whole antenne; those of the male insect are pectiniformed and bent so as to resemble the letter S in shape.

The same author has arranged such of the Linnæan Meloes as have the thorax scabrous, or rough, along with his *Cantharides*, and has preserved the generical name of our author to the *Mel. proscarabæus, No.* 1. This insect he
was

was obliged to separate from the others, in order to place it in his second section, or Coleopterous insects with elytra shorter than the abdomen.

All the Linnæan Meloes have five articulations in the tarsi of the two first, and four in those of the last pair of feet.

The larvæ of the Meloes feed chiefly on the leaves of plants, on which the compleat insects are likewise to be met with.

The insect called the Spanish Fly, or Blister-Beetle, belongs to this genus, though placed by Geoffroy among his Cantharides.

GENUS

Genus XXVIII. Mordella.

Linn. Syft. Nat. Pag. 682.

The antennæ of the Meloe are filiform, and ferrated.

The head is deflected, or bent under the neck.

The elytra are curve, or inclined downwards towards their point.

The palpi are compreffed, clubbed, and obliquely truncated.

A broad lamina is feated at the bafe of the abdomen, before the thighs.

Schæffer, defcribing the Mordel. Aculeata, Linn. No. 2, fays, that the thorax of that infect, and of the other Mordellæ, is convex, and narrower in the fore part than behind, and that the elytra are convex and margined; which obfervations hold good in all the infects belonging to this genus, which I have feen. According to the fame Author, their feet are faltatorii, or made for leaping.

According to Geoffroy, the antennæ of the Mordella are compofed of triangular articulations.

The tarfi of the firft pair of feet confift of four, and thofe of the laft pair of five joints.

The Mordellæ are common on flowers; their larvæ are yet unknown.

Genus XXIX. Staphilinus.

Linn. Syft. Nat. Pag. 683.

The antennæ of the Staphilinus are moniliform.

The elytra are not above half the length of the abdomen.

The wings are folded up, and concealed under the elytra.

The tail, or extremity of the abdomen, is *simple*, not being armed as that of the following genus, but is provided with two oblong veficles, which the infect can fhoot out or retract at pleafure.

Geoffroy differs from our Author, and from Scopoli, with regard to the antennæ of this genus, which, according to him, are filiform.

The tarfi, in all the feet, are compofed of five articulations.

The Staphilini are very voracious, devouring every kind of weaker infects, even thofe of their own fpecies. Some of them are found upon flowers,

flowers, but they chiefly inhabit the dung of cows: their larvæ which resemble them so much as scarce to be distinguishable, live in humid places under the ground.

The Staphilini are by some called *Rove-Beetles*.

Genus XXX. Forficula.

Linn. Syft. Nat. Pag. 683.

The antennæ in this genus are fetaceous.

The elytra are much fhorter than the abdomen,

The wings are folded, and covered by the elytra.

The extremity of the abdomen is armed with a kind of forceps, in which, and in the formation of the antennæ, this genus differs from the Staphilinus.

According to Schæffer, the wings of the Forficula are not entirely covered by the elytra, from under which I have frequently obferved the points to project.

The tarfi, in each of the feet, confift only of three articulations.

This infect is found every where in the fields, woods, and gardens, and is even at this time formidable to many people, from the idea that

it

it enters the ears, and pierces into the brain, which, however, anatomists know to be impossible, there being no communication between those parts, and the jaws of the insect too weak to effect one. It has been, from that circumstance, called the *Earwig*; the larva differs very little from the compleat insect, and is very lively, running with great agility.

ORDER II.

INSECTA HEMIPTERA.

The mouth and proboscis of the insects which compose this order, are inflected and bent inwards towards the breast.

The wings are *hemelytratæ*, or of a substance less hard and strong than those of the preceding order, but more so than the membranaceous ones of the following orders; the upper wings are semi-coriaceous; they do not meet together in a longitudinal suture, as in the foregoing order, but have some part of their interior margins crossed, or laid one over the other, above the abdomen.

This order contains the following genera, viz.

Genus I. Blatta.

Linn. Syst. Nat. Pag. 687.

The head of the Blatta is inflected.

The antennæ are setaceous.

The elytra and wings are extended, smooth, and semi coriaceous, or of a substance somewhat like vellum.

The thorax is rather flat, of an orbicular form, and margined.

The feet are cursorii, or made for running.

The abdomen is terminated by two little appendices, like horns.

To the above characters of the Blatta we may add, that the mouth is armed with jaws, and furnished with palpi; that the antennæ in most subjects are as long as the body, and that the abdomen is as broad as the thorax.

The upper wings cross over one another, above the abdomen, and are much stronger than the under ones, which last, according to Schæffer, are folded; in some subjects, however, they are extended like the elytra.

The tarsi of the fore feet have five joints, those of the hindmost have only four.

Geoffroy and Schæffer observe, that the horns which terminate the abdomen of the Blatta, are wrinkled or furrowed transversely.

The Blatta avoid the light, and with their larvæ, feed upon all kinds of food, but are more particularly fond of bread, meal, putrid bodies, and roots of plants; they are frequent in bakers shops, and in cellars; they fly the approach of danger with great swiftness; with us they are called *Cockroaches*.

The insect, called the *Kakkerlac*, so well known, and so much dreaded by the inhabitants of the American Islands, belongs to this genus.

Genus II. Mantis.

Linn. Syst. Nat. page 689.

The head of the Mantis is unsteady, or appears, from its continual nodding motion, to be slightly attached to the thorax. The mouth is armed with jaws and furnished with palpi.

The antennæ are setaceous.

The four wings are membranaceous, and wrapped round the body; the under ones are folded.

The anterior, or first pair of feet, are compressed, armed on the under side, with teeth like a saw, and terminated by a single nail or crotchet. The four hind ones are *gressorii*, or formed rather for advancing slowly, than for performing quick movements.

The thorax is extended to a considerable length, narrow, and throughout of equal size.

Scopoli

Scopoli has confounded this genus with the Gryllus as Linnæus had done in the tenth edition of his Syſtema Naturæ. It differs chiefly from that inſect in the number of articulations of which its tarſi are compoſed; (theſe in the Mantis are always five, but in the different families of Grylli, are ſometimes three, ſometimes four) and in its having only one crotchet or nail, to thoſe of the firſt pair of feet.

The eyes of the Mantis are prominent, and its head perfectly reſembles thoſe of the ſecond family of the Linnæan Libellulæ.

The elytra are not much ſtronger than the under wings.

The abdomen is terminated by little appendices or horns, leſs ſtiff than thoſe of the Blattæ; that part is not always long and narrow, as aſſerted by Schæffer, but in ſome ſubjects flat and very broad compared with its length. The laſt mentioned author calls the feet *ſaltatorii* made for leaping, which they do not appear, nor are obſerved to be, by any other author I have met with.

This inſect is, with us, called the *Camel Cricket*. It is looked upon by the Africans as a ſacred

cred animal (according to Geoffroy, the French peasants hold it nearly in the same light), from its frequently assuming a praying or supplicating posture, resting upon its hind feet, and elevating and folding the first pair.

Genus VI. Gryllus.

Linn. Syst. Nat. page 692.

The head of the Gryllus is inflected, armed with jaws, and furnished with palpi.

The antennæ, in some subjects, are setaceous, in others, filiform.

The wings are declined towards, and wrapped round the sides of the body; the under ones are folded up, so as to be concealed under the elytra.

All the feet are armed with two nails or two crotchets; the hind ones are formed leaping.

The Grylli are divided into different sections as follows:

1. The *Acridæ,* which have the head of a conic form, and longer than the thorax; their antennæ are ensiform, or somewhat resembling a sword.

2. *Bullæ*

2. *Bullæ*, which are diftinguifhed by a kind of creft or elevation on the thorax: Their antennæ are fhorter than the thorax, and filiform.

3. *Achetæ*, which are known by two *Setæ* or *Briftles*, fituate above the extremity of their abdomen. The houfe-cricket belongs to this family.

4. *Tetigoniæ*: The females in this fection are diftinguifhed by a kind of tube with which the extremity of their abdomen is furnifhed, and through which they depofit their eggs in the ground. The antennæ in both fexes of this family are fetaceous.

5. *Locuftæ*, in which fection the tail is fimple, without the fetæ by which the *Achetæ* are diftinguifhed, or the tube that terminates the tail of the females in the preceeding genus. Their antennæ are filiform,

Geoffroy

Geoffroy has formed from some of these sections as many different genera.

To the *Achetæ* of our author he attributes the generical name of *Gryllus*, adding to the Linnæan characters, that they have three stemmata, and that the tarsi are composed of three articulations.

To the *Locustæ* he has given that of *Acrydium*, adding, that the antennæ are one half shorter than the abdomen, that they have three stemmata, and three joints to the tarsi, as in the last mentioned genus. And

To the *Tetigoniæ*, that of *Locusta*, these, according to him have filiform antennæ longer than the abdomen, and differ from the two preceeding genera in the formation of their tarsi, which have four articulations. Schæffer has followed him in this disposition of the Linnæan Grylli, each having first arranged them in different orders, according to their own system. He observes, that the upper wings of each genus are less transparent, but of a stronger substance than the under ones.

The larvæ, or caterpillars of the Grylli, very much resemble the perfect insects, and, in general, live under ground. The Chrysalids very much

much resemble and accompany their parents, many of which feed upon the leaves of plants. Others, which live in houses, prefer bread, meal, and every kind of farinaceous substance; some of them are with us called *locusts*, others *grasshoppers*, others again, *Crickets*.

Genus IV. Fulgora.

Linn. Syst. Nat. page 703.

The front, or fore part of the head of the Fulgora is drawn out, extended, and empty.

The antennæ are seated below the eyes, having two articulations, whereof the exterior is larger, and of a globular form.

The rostrum is inflected, or bent inwards under the body.

The feet are formed for walking. In this circumstance particularly it differs from the following genus, with which it was confounded before the last edition of the Syst. Naturæ.

This genus seems to have been unknown to Geoffroy, Schæffer, and Scopoli. One of the insects belonging to it is however found in Germany, and two different species have been caught in this country; the one by the author of that useful and elegant work the Flora Anglica, the other by my friend Mr. Grey. Whether the

larvæ

larvæ of thofe infects (which differ very little from fome fpecies of the Cicadæ) refemble thofe belonging to that genus or not, is yet unknown.

The foreheads of many Fulgoræ (efpecially thofe found in China) emit a very lively, shining light, in the night-time, which, according to fome authors, is fufficient to read by; I have not heard that the European fulgoræ poffefs that quality.

GENUS

Genus V. Cicada.

Linn. Syſt. Nat. page 704.

The roſtrum of the Cicada is bent inwards, under the breaſt.

The antennæ are ſetaceous.

The four wings are membranaceous, declining along the ſides of the body.

The feet in moſt ſubjects are formed for leaping, in others (particularly the *manifera*) for walking or creeping.

They are divided into different ſections, as follow:

1. The *Foliaciæ*, in which the thorax is compreſſed, membranaceous, and larger than the body.

2. The *Cruciatæ*, which have the thorax armed on each ſide with a horn, or ſpine.

3. The *Maniferæ*, distinguished by their feet, which are not made for leaping.

4. The *Ranatræ*, which differ from the last section in their hindmost feet, which are saltatorii, or made for leaping.

5. The *Deflexæ*, whose wings are wrapped round the sides of the body.

Geoffroy observes, that the antennæ of the Cicada are shorter than their head, and that the under wings are crossed one over the other.

Scopoli has divided the Cicadæ into different sections, from the substance of their elytra; the first having those parts wholly coriaceous; in the second, they are coriaceous only half their length; in the third, they are membranaceous.

The pupæ, or chrysalids, of many Cicadæ, differ from the perfect insect only in the shortness of their elytra and wings; they run and leap upon plants and flowers with great agility. The larvæ of the *Ranatræ* discharge a kind of froth from the anus and pores of the body, under which they conceal themselves from the rapacity

rapacity of such stronger insects as prey upon them. Those of the Maniferæ pass a whole year under ground; these last make a noise like the cricket.

The Cicada is called by some English authors, the *Frog-hopper*; by others, the *Flea-locust*.

Genus VII. Notonecta.

Linn. Syst. Nat. page 705.

The rostrum of the Notonecta is inflected.

The antennæ are shorter than the thorax.

The four wings which are coriaceous from their base to their middle, are folded together crosswise.

The hind feet are hairy, and formed for swimming.

Geoffroy adds to the above characters of the Notonecta, that it has an escutcheon, that its tarsi have two articulations, and that all the six feet are equally formed for swimming, which they appear to be in all the Linnæan species, except the *Not. Striata*, Linn. Syst. Nat. No. 2. From this insect Geoffroy has formed a separate genus termed *Corixa*, with the following distinct characters:

No escutcheon.

The tarsi containing only of *one* articulation.

Six feet, the anterior pair heliform, or like the claws of a crab, the laſt pair only formed for ſwimming.

In this he is followed by Schæffer.

The Notonectæ are not uncommon in ſtanding waters; they ſwim upon their backs on the ſurface of the water with great agility; their larvæ reſemble them very much. The name of *Boat-fly* has been given them, not improperly, by ſome Engliſh authors.

The abdomen of the Notonecta is terminated by four little horns or appendices.

Genus VII. Nepa.

Linn. Syft. Nat. page 118.

The roftrum of the Nepa is bent inwards.

The antennæ   —   —

The four wings are folded together croffwife, with the anterior part coriaceous as in the preceeding genus.

The two fore feet are cheliform, or refemble the claws of a crab; the four others are formed for walking.

Geoffroy afferts that the Pedes Cheliformes, or fore feet of Linnæus, are the antennæ of the infect, which according to him has but four feet. That author has given to this genus the name of *Hepa*, and adds that the Tarfi are compofed of one fingle articulation.

He has formed a diftinct genus from the *Nepa Cimicoides* of Linnæus, in which infect he had difcovered very fhort antennæ fituate under the eye; and which is farther diftinguifhed from the other Nepæ, by having

ing tarsi composed of two articulations. This genus he has named *Naucoris*. Schæffer has pursued the same method preferable to that of our author, who is followed by Scopoli.

The last mentioned author has observed, by the help of the microscope, a tubercule, or small elevation, near the eyes of the Nepa, on which are two or three hairs, which he takes to be the antennæ.

The Nepæ are well known by the name of *Water Scorpions*. They are frequent in standing waters, as well as their larvæ and chrysalids, both which resemble them very much. They live chiefly upon aquatic insects, and are exceedingly voracious.

Genus VIII. Cimex.

Linn. Syst. Nat. page 715.

The rostrum of the Cimex is inflected.

The antennæ are longer than the thorax.

The wings are folded together crosswise; the upper ones are coriaceous from their base towards their middle.

Their back is flat.

The thorax is margined.

The feet are formed for running.

This genus is divided into different sections, as follows:

1. The *Apteri*, or those without wings.

2. The *Scutellati*, in which the escutcheon is extended so far as to cover the abdomen and the wings.

3. The *Coleoptrati*, whose elytra are wholly coriaceous, not becoming membranaceous towards their extremity, as in the other Cimices.

4. The *Membranacei*, whose elytra are membranaceous; these are very much depressed, like a leaf.

5. The *Spinosi*, in which the thorax is armed, on each side, with a spine.

6. The *Rotundati*, which are of an oval form, without spines on the thorax.

7. The *Seticornes*, in which the antennæ become setaceous towards their point.

8. The *Oblongi*, or those of an oblong figure.

9. Those whose antennæ are wholly setaceous, and as long as the body.

10. The *Spinipedes*, which have their thighs armed with spines.

11. The *Lineares*, distinguished by their long and narrow body.

Geoffroy observes, that the antennæ of the *Cimices* are composed either of four or five articulations (from which circumstances he has divided them into two families) and that they are longer than the head.

The tarsi have five articulations.

The larvæ of the Cimices run about, and, like the compleat insect, suck in their food through their beak: many of them live upon the juices of plants, others upon the blood of animals; several are found in the waters, and others frequent houses, among which is the common Bed-Bug, an insect but too well known. They differ from other insects in their softness, and most of them emit a very fœtid smell.

The common Bed-Bug belongs to the family of Apterous Cimices. Scopoli, however, pretends that it is likewise found with wings.

Genus IX. Aphis.

Linn. Syst. Nat. page 733.

The rostrum of the Aphis is bent inwards.

Their antennæ are longer than the thorax.

They have either four erect wings, or are without wings.

Their feet are made for walking.

They have generally two little horns or spines placed on the hinder part of their abdomen.

Schæffer asserts, that all the *male Aphides* have wings, and that all the females are apterous.

The tarsi, in each sex, have only one articulation.

The antennæ are setaceous.

Geoffroy has observed, that the aphides have two beaks, one of which is seated in the breast, the other in the head; this last extends to, and is laid upon the base of the pectoral one,

and

ferves, as that author fuppofes, to convey to the head a part of that nourifhment which the infect takes or fucks in, by means of the pectoral beak.

The infects belonging to this fingular genus, in the fummer bring forth live young, and in the autumn lay eggs. Entomologifts affert, that from the copulation of the parents fpring daughters, grand daughters, great-grand-daughters, and great great-grand-daughters, or females fœcundated to the fifth (according to *Bonnet*, to the ninth) generation, fome with, others without wings, without diftinction of fex, in the fame fpecies; many of them are provided with two horns on the hinder part of the abdomen, with which they extract the fweet-tafted dew from flowers.

The Aphides are devoured by the larva of the *Myrmelion Formicarium* of Linnæus; Ants are likewife very fond of them, on account of a fweet liquor with which their bodies are humected. They are exceeding common, and are generally termed the *lice* of the plant which each particular fpecies infeft.

Genus X. Chermes.

Linn. Syft. Nat. page 377.

The roftrum of the Chermes is placed in the breaft.

The antennæ are longer than the thorax.

The wings are declined along the fides of the abdomen.

The thorax is gibbous.

The feet are made for leaping.

Geoffroy has named this genus *Pfylla*, and obferves that its abdomen ends in a point, that it has three ftemmata, that the roftrum is fituate between the firft and fecond pair of legs, and that the tarfi are compofed of two articulations.

Schæffer, who with Scopoli has preferved the Linnæan name to this genus, fays, that the antennæ are fetaceous, and longer than the thorax. The larvæ of the Chermes have fix feet, refemble the compleat infect, and are generally covered with a hairy or woolly fubftance. The winged

winged insects leap or spring with great agility, and infest a great number of different trees and plants: the females insert their eggs under the surface of the leaves, by means of a tube, with which their abdomen is armed, thereby causing the little tubercules, or galls, with which the leaves of the ash, the fir, and other trees, are sometimes almost wholly covered.

Genus

Genus XI. Coccus.

Linn. Syst. Nat. page 569.

The rostrum of the Coccus is situate in the breast.

The hinder part of the abdomen is bristly.

The males have two erect wings.

The females are apterous.

Schæffer observes, that their antennæ are setaceous.

The female Cocci fix themselves and adhere, almost immovably, to the roots, and sometimes to the branches of plants, where they are visited by the winged males; some of them having thus fixed themselves, lose entirely the form and appearance of insects; their bodies swell, their skin stretches, and becomes smooth, the segments of their abdomen disappear, and they much resemble some kinds of galls or excrescences found frequently on the leaves and branches of plants, that in general they are mistaken for such; after which changement, the abdomen serves only for a kind of covering or shell, under which the eggs are concealed; to these Geoffroy has

has given the generical name of *Chermes*. Others, again (though they likewise fix themselves, and adhere immovably to the leaves of plants, like Chermes) preserve the form of insect till they have laid their eggs and perish; to these, that author has preserved the Linnæan generical name of *Coccus*. These are likewise distinguished from the Chermes by the form of their abdomen, which part, in the females, is more oblong, and composed of a greater number of segments than in the females of the other genus; a kind of down, or cotton, likewise grows out of their belly, which serves as a nest in which they deposit their eggs; the males of all of them are much less than the females, and the larvæ of all the different species perfectly resemble one another.

These insects, whether the Linnæan method of arranging them, or that of Geoffroy is adopted, differ (as before observed) from all other Dipterous ones, in the want of *halteres* or *poisers*, and from the other classes, in the number of their wings, which circumstances render them very distinguishable.

GENUS

Genus XII. Thrips.

Linn. Syst. Nat. page 743.

The rostrum of the Thrips is obscure, or so small as to be scarce perceptible.

The antennæ are as long as the thorax.

The body is slender, and of equal thickness in its whole length.

The abdomen is reflexible, or bent upwards.

The four wings are extended, incumbent upon the back of the insect, narrow in proportion to their length, and cross one another at some distance from their base.

Geoffroy says, that the antennæ in this genus are filiform.

He has not been able to discover the proboscis of this insect, and asserts, with Schæffer, that the mouth is formed by a simple longitudinal cleft, in which, he adds, it is possible that the jaws may be concealed; and as the Thrips would, in

his

his opinion, be a Coleopterous insect, if those jaws really existed, he has taken that circumstance for granted, and has accordingly arranged it under that class: the other reasons for which he has assigned it that place, appear to me without force, since the characteristics from which he has deduced them are likewise found in the Hemipterous insects; these are the form of the antennæ, their position, that of the legs, the two first of which are attached to the thorax, the four others to the abdomen, and the consistence of their elytra, which are less flexible than the wings.

The tarsus of each foot has only two articulations, the second of which Bonani and others have observed to form a kind of vesicle, or bladder.

These insects are very common on flowers, upon which they run, or rather leap, with great vivacity, often bending their bodies upwards. Their habitation is generally under the bark of trees.

Scopoli has observed that they skip or spring rather by means of the abdomen, than of their feet; they are in general so small as scarce to be perceptible. Their larvæ run as briskly as themselves, and are often of a red colour.

ORDER

ORDER III.

INSECTA LEPIDOPTERA.

The insects which compose this order have four wings, covered with a farinaceous powder, or a kind of scales, disposed in regular rows, nearly in the same manner as tiles are laid upon the roofs of houses. The beautiful colours which adorn the wings of Lepidopterous insects are formed by these scales, and if, by any accident, they are rubbed off, the wings appear to be nothing more than a naked membrane.

Their mouth is furnished with a spiral tongue, which they can unfold or extend, and roll up again at pleasure.

Their bodies are hairy.

This order is divided into three genera, viz.

Genus I. Papilio.

Linn. Syst. Nat. page 744.

The antennæ of the Papilio (generically known with us by the name of *Butterfly*) grow thicker towards their extremity, and are in most subjects terminated by a kind of capitulum, or head.

Their wings, when sitting or at rest, are erect, insomuch, that their extremities meet or touch one another above the body.

They fly in the day-time.

They are divided into sections, distinguished one from another by the following characteristics.

1. The *Equites*, known by the shape of their superior or upper wings, which are longer from their hinder corner or angle to their anterior extremity, than from the same

same point to their base; some of these have filiform antennæ, in which particular they resemble the genus *Phalena*, or moths, from which, however, they are easily distinguished, by their outward appearance, their bodies being much lighter, or less bulky, and not so well covered with hair, and by the shape of their upper wings.

The Equites are either

Troes, which are known by the bloody spots found upon their breasts; these are likewise generally of a dark or black colour: Or

Achivi, on the breasts of which the bloody spots of the *Troes* are not found, and are farther distinguished by an ocellum, or spot, resembling an eye, situate at the inner corner of their posterior wings;

wings; the colours of the Achivi are generally gay, and and they are either

Simple, of one colour : Or

Variegated, adorned with various colours.

Such of the Equites as we are acquainted with have fix feet.

2. The *Heliconii* : thefe are diftinguifhed by the narrownefs of their wings, which fometimes appear (efpecially towards their extremities) to be naked, or deprived of fcales; their upper wings are of an oblong form, the under ones are very fhort in proportion to their breadth: this laft characteriftic, however, is not univerfal; fome infects, which refemble the *Heliconii* in every other particular, being referred to that fection, though their under wings are proportionably

portionably long; as the *Pap. Appolo, Mnemosyne,* &c. all their four wings have the edges or margins entire.

3. The *Danai,* the edges of whose wings are entire. They are either

> *Candidi,* the ground colour of whose wings is always white, or
>
> *Festivi,* the canvas, or reigning colour of whose wings is never *white;* these are likewise adorned with a great variety of colours, which seldom occurs in the *Candidi.*

The *Danai* resemble the *Heliconii* in the edges of their wings, being entire, but are easily distinguished by the shape of them, those of the Danai being round, those of the Heliconii oblong; they appear likewise to be of a stronger texture, and rougher, being better covered with scales, especially at their extremities.

4. The

4. The *Nymphales*, distinguished from the *Heliconii* and *Danai*, by the edges of their wings, which are indented or scolloped. They are either

Gemmati, in which family the wings are adorned with eyes; these eyes are found on all the four wings, in some species, in others on the upper wings, in others on the under wings only : Or

Phalerati, the wings in which division want the eyes by which the Gemmati are distinguished, but are not less beautiful, being generally painted with a great variety of colours.

5. The *Plebeii*, which are smaller in general than the others, and are either

Rurales, distinguished by the spots on their wings being *obscure*, which term does not regard the colours of the spots, often very beautiful, but their nature, they not being pellucid, or transparent : Or

Urbicoli, the spots on the wings of which are for the most part transparent.

The division of the Butterflies into families, from the circumstances chosen by Linnæus, seems liable to many objections; the family of the Plebeii, in particular, is very inaccurate, and contains insects very different from one another, at the same time that they resemble, and have all the characters of some or other of the preceding ones, under which many of them, I think, might be properly arranged; the remaining Plebeii would compose a family very distinct from all the others, and which might be formed into two sections, the first containing small Butterflies, having long and flexible or weak tails, slender bodies, and clubbed antennæ, as the *Cupido*, the *Marsyas*, the *Bæticus*, &c. the other distinguished by the shortness, thickness, or breadth of their head, thorax, and abdomen, and by the shape of their upper wings, which in these last are pointed at their extremity, and long in

proportion

proportion to their width, as the *Proteus, Phidias*, and others.

The antennæ in this last division are generally uncinnated or crooked at their extremity; some of them have likewise tails, but these are very broad and strong, and are always ciliated or edged with a fringe of hairs, as in the *Proteus*, &c.

The bloody spots mentioned by Linnæus, are not always found on the breasts of the Eq. Trojani, nor is the interior angle of the Achivi always adorned with an eye, so that the surest method is to refer such Equites as are of dark or mourning colours, to the family of the *Troes*, and those of gay, lively ones, to that of the *Achivi*.

The under wings of a great many of the Papiliones, placed by Linnæus among his *Heliconii*, are slightly indented, and as they are without eyes, they ought, strictly speaking, to be referred to the *Nymphales Phalerati*, but are distinguishable by the delicacy of their texture, and the smoothness of their wings, which are less covered with scales than those of the last-mentioned family.

The under wings of the *Danai Festivi* are likewise often indented, but in that case they are generally edged with a kind of fringe, or their margins, especially on the under side, surrounded by one or more *waved* lines, or rows of white spots; those Butterflies, therefore, whose wings are but *slightly* indented, adorned with eyes, and the margins surrounded by rows of white spots, or *narrow, waving* lines, belong rather to this family than to that of the *Nymphales Gemmati*.

Scopoli and Geoffroy have divided this genus into different families principally from the number of their feet; a method which cannot easily be pursued in cabinets where exotic Butterflies are admitted, these parts being generally destroyed before such insects reach Europe. The other circumstances from which Geoffroy has taken his divisions into families, viz. the form of the caterpillars, is totally impracticable, except where the collector admits no other Butterflies into his cabinet but such as he has himself possessed in the caterpillar state.

The pupæ of all Butterflies are *obtectæ* and *naked*, and, except those of the *Danai Candidi*, are suspended *perpendicularly* in the open air, being attached by their tail to the under sides of branches of trees, leaves of plants, &c.

Thofe of the Danai Candidi (at leaft of fuch as we are acquainted with) are fufpended *horizontally*, being fixed by the tail as thofe of the other families, but are fupported in an horizontal pofition by means of a thread paffed round the middle of their body and attached obliquely to the part above the head.

The caterpillars of many of them are exceedingly common, and fufficiently known; thofe of many *Equites* have two horns fituate in their necks, near the head, which they can fhoot out and draw in at pleafure. It is yet unknown, whether or no the others of that fection have thefe horns, but it is to be hoped that fome curious Entomologift will make this point an object of his refearches: the larvæ of the Pap. *Apollo* refembles thofe of the Equites in that refpect.

Genus II. Sphinx.

Linn. Syft. Nat. page 796.

The antennæ of the Sphinges are thicker in the middle than at the extremities, somewhat refembling a prifm in form.

The wings are deflected, the outer margins declining towards the fides.

Their flight is flow and heavy.

They are divided into families, as follows:

1. The *Legitimæ*: thefe have either

    Angulated wings, with the *anus* fimple, not terminated by a tuft of hair: Or

    Entire wings, with the anus fimple: Or

    Entire wings, with the anus terminated by a tuft of hair.

2. The *Adfcitæ*, differing from the others in their external appearance and Caterpillars.

The Sphinges fly either early in the morning, or after sun-set in the evening; they for the most part were heavily and sluggishly, often emitting a kind of sound. They suck the nectar of flowers with their tongues, though they seldom settle upon them: most of them undergo their metamorphosis in the earth; their chrysalids are *obtectæ*, but inclosed in a kind of covering, or web, composed generally of course materials, in which particular they differ entirely from the preceding genus, the chrysalids belonging to which are naked, and suspended in the open air.

The bodies of most of their caterpillars are smooth, or without hair, and have a horn or spine situate above the anus; that, however, of the *Sphinx Filipendulæ Linn. No.* 34, wants this horn, as Geoffroy has observed; for which reason that author has separated it from the other species, that insect alone composing his third family: his two others are distinguished by their having or wanting tongues; the antennæ of these two last-mentioned families, he says, are prismatic, but throughout of equal thickness, those of the Sph. Filipendulæ, on the contrary, are much larger in the middle than towards their extremities.

Scopoli

Scopoli has divided the Sphinges into two sections; the first containing such as undergo their metamorphosis in the ground; the second, those which undergo their last changement above ground. This method can only be pursued by those who observe the metamorphosis of every Sphinx they place in their collection, since it is impossible to procure its natural history along with every insect, especially such as are sent from far distant countries: the divisions of genera into sections should always be taken from some remarkable circumstance found constantly upon the insects, after their death.

The name of *Hawk-moth* has been given, by most English authors, to the Sphinx.

GENUS III. PHALENA the MOTH.

LINN. Syft. Nat. page 808.

The antennæ of the Phalenæ are fetaceous, decreasing in fize from the bafe towards the point.

Their wings, when at reft, are in general deflected.

They fly in the night.

This genus is divided into the following families:

1. *Attaci*, whofe wings incline downwards, and are fpread open.

    Thefe have either

        Pectinated antennæ, without a tongue,

        Pectinated antennæ, and a fpiral tongue: Or

        Setaceous antennæ, with a fpiral tongue.

2. *Bombyces*,

2. *Bombyces*, whose wings cover the body in a position nearly horizontal, and which have pectinated antennæ.

These are either

Elingues, which want the tongue, or have it so short as not to be manifestly spiral.

Again their wings are either

Reversed or *deflected*.

Or *Spirilingues*, which have a spiral tongue, and are either

Cristatæ, with a kind of crest, or tuft of hair on the back, Or

Læves, with smooth backs.

3. *Noctuæ*, whose wings are incumbent, as in the Bombyces, from which they differ chiefly in the formation of their antennæ, which are setaceous.

The

The Noctuæ are either

> *Elingues,* wanting tongues ; Or
>
> *Spirilingues,* having spiral tongues.

4. *Geometræ,* whose wings, when at rest, are extended horizontally.

The antennæ, in one subdivision of this section, are

> *Pectinated*

In another,

> *Setaceous.*

The under wings in each of these divisions are either

> *Angulated,*
>
> Or *round,* with entire edges.

5. *Tortrices:* the wings of the Tortrices are exceedingly obtuse; their exterior margin is curve, and declines

lines towards the sides of the body.

These have short palpi.

6. *Pyralis:* the inner margin of the wings in this section are laid one over the other; the wings themselves decline a little towards the sides of the body, and in shape resemble a *delta,* or *triangle*; these have considerable palpi of different forms, which has induced Scopoli to divide them into two sections; the one containing those whose palpi are curve, or bent upwards; the other, those in which these parts are extended.

7. *Tiniæ:* the wings of the Tiniæ are wrapped or folded up round the body so as to give the insect a cylindrical form; the forehead is stretched out, or advanced forwards.

Many of the Tiniæ have incumbent wings expanded their whole breadth, and seem to form a very distinct section, differing from the Tiniæ in that particular; from the Pyralides in the want of palpi, and distinguished from the other families of Phalænæ by their porrected forehead, and a kind of fringe, with which the interior margins of their wings, are edged.

8. *Alucitæ*: the wings of this division are split, or divided into branches, almost to their base.

Geoffroy has separated the last family of our author, the *Ph. Alucitæ*, from the other Phalœnæ, under the generical name of *Pterophorus*, on account of the chrysalids of the insects belonging to it being naked, and suspended horizontally in the open air, as those of the *Danai Candidi*, or third family of butterflies, in which particular they certainly differ essentially from the Phalœna, whose Chrysalids are either concealed in the ground, or protected from the inclemency of the weather by a covering, which
some

some of them, as the Silk-worm, compose of the richest materials.

That author has likewise formed the *Tiniæ* into a separate genus with the Linnæan characters and name. The remaining Phalenæ he has divided into two families; in the first of which the antennæ are pectinated; in the other filiform; these families again are subdivided into the *Elingues*, and the *Linguatæ*, in each of which sections the wings are, in some subjects, *deflected*, in others, *extended horizontally*; among these he has dispersed the *Tortrices* and *Pyralides* of Linnæus.

Scopoli observes that this division, taken from the antennæ, labours under very great difficulties, those parts being formed differently in the different sexes of Phalenæ, besides that of procuring both sexes of each species compleat; for these reasons he has reunited the Bombyces and Noctuæ of Linnæus (separated by Geoffroy on account of the different formation of their antennæ) under the title *Bombyx*; these he has divided into two sections, the insects belonging to one of which, undergo their transformation under the ground; the others above ground; the tongue is wanting in the first of these

these sections; in the other, some have, some want, that part.

The Geometræ are divided by him into three sections; the first having angulated, the second dentated, the other entire wings.

His division of the *Pyralides*, taken from the palpi, has been mentioned above.

His *Tineæ* are likewise formed into two sections, from their wings being *convoluted* or *extended*.

The caterpillars of moths are either

Smooth, without elevations,

Or *tuberculated*, with small gibbosities upon their bodies, resembling knots.

Naked, without hairs or down,

Or *hairy*.

They differ likewise in the number of their feet, as follows:

The *Bombyces* and *Noctuæ* have sixteen feet.

The *Tineæ* have fourteen.

The *Phalæna Gamma* alone has twelve.

Moſt of the *Geometræ* have ten.

Thoſe of all the different families have ſix feet at the breaſt, or fore-part of the body, except the *Vinula*, *Furca*, *Lacertina*, and a few others; the chief difference, therefore, lies in the abdominal feet, which are either eight, ſix, four, or two, in number, or are entirely wanting.

The caterpillars of the Geometræ have ſix pectoral or fore feet, two tail, or hind feet, two others, a little before the tail, and want the abdominal ones, which makes them reſemble Leaches in their gait. From the ſame circumſtance, likewiſe, the name of *Geometræ* has been given to them, becauſe they ſeem to meaſure the ground over which they advance. They reſt in an erect poſture, ſupported only by the feet ſituate under their tail: thoſe of ſuch of this family as have pectinated antennæ, reſemble ſo much the branches of the plants upon which they feed, as not to be eaſily diſtinguiſhed from them. This reſemblance, without doubt, contributes very much to preſerve them from the voracity of the different birds which prey upon them.

The

The caterpillars of the *Tortrices* roll up, and fasten together by a thread the leaves of the plants upon which they feed; thus securing to themselves a kind of retreat.

The caterpillars of most of the *Tineæ* keep always under some kind of covering, where they live and feed in security; some of them roll up the leaves of plants for their habitation; others, which feed only upon the interior surface of leaves, lodge themselves under the epidermis, or exterior skin; others, again, in woollen cloths, skins of beasts and birds, &c. These all undergo their metamorphosis in the places and under the coverings in which they had lived; some few live in society under a kind of web formed by their joint industry: the moths which are produced from these last have generally, as Linnæus observes, expanded wings.

According to Geoffroy, the caterpillars of some Tiniæ have eight, others have fourteen, and others, again, sixteen feet.

The pupæ, or chrysalids, are either simple, or have a kind of hook at their extremity; they are all enclosed in a web or covering; some of them pass that state under the ground, others are fixed to the under sides of branches of trees, walls, &c. The webs of the first consist generally of very coarse materials, strongly attached together by a few threads of silk. Those of the others have generally more silk, and are weaved more naturally; that of the *Silk-worm* furnishes an article which long was considered as for mere luxury, but which is now of universal utility.

ORDER

ORDER IV.

INSECTA NEUROPTERA.

The insects belonging to this order have four membranaceous, naked wings, reticulated with veins, or in which the membranes cross one another so as to appear like net-work.

Their tail is unarmed, or without a sting, but is frequently furnished with appendices like pincers, by which the males are distinguished.

This order contains the following genera:

Genus I. Libellula.

Linn. Syst. Nat. Pag. 901.

The mouth of the Libellula is armed with jaws, which are always more than two in number.

The antennæ are shorter than the thorax.

The wings are expanded, without folds.

The tail of the male is furnished with a kind of forceps.

The libellulæ are divided into two families; the first

With wings extended horizontally, when at rest; the second

Distinguished by the eyes being placed at a distance from one another; the wings in this family are erect and the eyes very prominent.

Geoffroy adds to the above characters of the Libellulæ, that they have three stemmata situate between the eyes, and that their tarsi are composed of three articulations. He divides them into two sections; the first having erect wings (when at rest) the second *patent*, or *open* and *extended wings*.

The insects belonging to the first division of Libellulæ live chiefly upon Moths, the others upon Muscæ, or flies; they are all exceedingly ly voracious; Linnæus calls them the Hawks of Gymnopterous infects. The larvæ of both live and run, rather than swim, in the water; they devour aquatic insects weaker than themselves, and are not less voracious than the compleat insects; they are likewise exceedingly cruel, being frequently observed to kill and tear other insects to pieces when not pressed by hunger, since they leave the carcases entire.

The figure of the larva is very singular, and may be seen in *Geoffroy, tom.* 2. *tab.*

The chrysalis differs very little from the larva, and like it runs with great agility in the water, devouring smaller insects. It generally quits the water before it undergoes its final changement.

The manner in which some of the Libellulæ effectuate the work of generation is truly singular: the male pursues his female on the wing, and instead of endeavouring to win her by gentle means to his embraces, seizes her with the forceps at his tail by the neck, where he holds her fast, till she, to get quit of so cumbersome a burthen, willingly, or unwillingly, approaches her tail, in which are situate her organs of generation to the breast of her ravisher (under which his sexual parts are placed) thus united in a kind of ring, the male not quitting his hold of the female's neck, they continue their flight until the work is performed.

The Libellulæ are by some called *Dragon-flies*, by others, *Adder-bolts*; they are frequently met with in the summer season, near standing waters, where the females go to deposit their eggs; the different sexes are often differently coloured, and the species vary very much, which renders it difficult to distinguish them.

Genus IJ. Ephemera.

Linn. Syft. Nat. page 906.

The mouth of the Ephemera has neither teeth nor palpi.

The ftemmata are two in number, fituate above the eyes, and larger than they are generally found to be in other infects,

The wings are erect, the hinder ones much fhorter than the others.

The tail is furnifhed with appendices, refembling hairs, or briftles.

They are divided into two fections; the firft having two, the other three fetæ, or briftles, on the tail.

Geoffroy afferts that the ftemmata are three in number, which I have obferved them to be in fome fpecies.

Schæffer adds to the Linnæan characters of the Ephemera,

That

That the antennæ are fetaceous and fhort.

That the tarfus of each foot has five articulations, and

That the thorax is very fhort.

Their flight is flow and heavy, which renders them an eafy prey to fwallows, and other birds.

Thefe infects differ in many particulars from all others; their caterpillars live in the water, where earth and clay feem to be their only nourifhment for three whole years, the time they confume in preparing for their metamorphofis, which they undertake and effect in a few moments. The larva, when ready to quit that ftate, arifes to the furface of the water, and getting inftantaneoufly rid of his fkin, becomes a chryfalis. This chryfalis is furnifhed with wings, which it makes ufe of to fly to the firft tree or wall it meets, and there fettling, in the fame moment quits a fecond fkin, and becomes a perfect Ephemera. In that ftate, for which it had been fo long preparing, the pleafures it enjoys muft be very fenfible, if they are lively in proportion to the fhortnefs of their duration; the

infect

insect generally celebrating its nuptials, producing the fruit of them, and dying within the space of a few hours, seldom or never surviving the day on which it may be said to have really begun to live.

They differ no less from other insects, in the manner of propagating their species, than in the shortness of their lives, and their long continuation in the caterpillar state. The female Ephemera has no sooner quitted her chrysalis, than she returns to the water from whence she sprang, upon the surface of which she lays her eggs; the male, attentive to all her motions, takes care immediately to fecundate the eggs, nearly in the same manner as fish fæcundate those of their females. (Geoff.)

The antennæ of the perfect insect resemble hairs, being without joints or articulations. When at rest, the fore legs are advanced or stretched out before the head.

The Ephemeræ are very frequent near waters: they multiply amazingly in some places, insomuch that Scopoli asserts the peasants in his neighbourhood to be discontented with
their

their share of them, unless each can collect at least twenty cart-loads, making use of them for manuring their lands, which purpose they answer exceedingly well. They are called with us *May flies*.

Genus III. Phryganea.

Linn. Syſt. Nat. page 908.

The mouth of the Phryganea is without teeth, but is furniſhed with four palpi.

The ſtemmata are three in number.

The antennæ are longer than the thorax.

The wings are incumbent, or laid horizontally on the body.

The under wings are folded, ſo as to be concealed under the upper ones.

This genus is divided into two ſections.

The firſt diſtinguiſhed by two truncated ſetæ, reſembling unſpun ſilken threads, which terminate the abdomen.

In the ſecond the abdomen is ſimple, or wants thoſe appendices.

Geoffroy has ſeparated theſe two families of Phryganeæ,

Phryganeæ, and given to the first the generical name of *perla*. These *perlæ* differ from the other Phryganeæ (to which he has preserved that generical name) not only in the appendices of the tail, but also in the position of the wings, which, in the latter, decline from the inner margins, towards the sides, so as to resemble the ridge of a house, and are curve, or turned upwards, at their extremity; and in the number of articulations, which compose their tarsi, these, in the Perla, are but three; in the Phryganea they are five.

The Perlæ and Phryganeæ, however, do not seem to differ generically; their larva perfectly resembling one another, and their manner of living the same; they likewise perform their metamorphosis in the same season, and in the tubes in which they dwell while larvæ. The latter, however, remain considerably longer in the chrysalis than the Perlæ. In the year 1768, I had an opportunity of observing the metamorphosis of three of the perlæ, and four phryganea; the Chrysalids were all kept together, and in the same degree of heat: two of the perfect insects were produced on the eighth, and another on the ninth day (after their respective

transfor-

transformations); thefe three proved all to belong to the firft family of Linnæan Phryganeæ, or the Perlæ of Geoffroy. Another perfect infect quitted the chryfalis on the fourteenth day after its entering into that ftate, and two others on the nineteenth day: the three laft proved all to be Phryganeæ, of the fecond Linnæan divifion, or the Phryganeæ of Geoffroy; the other chryfalis perifhed without coming to perfection. This circumftance, however, will, I prefume, fcarce be deemed fufficient to form a generical diftinction between the two infects, tho' when added to the others before-mentioned, they may jointly render the divifion of the Genus into families, very proper.

Scopoli has preferved the Linnæan Genus intire, with the fame characters, as that author has affigned to them, but has taken his divifion of it into families, from different circumftances. In his firft, the wings are incumbent, in the other deflected. That author has obferved, that one fpecies of the lezard is exceedingly fond of the Phryganea, and that the Phryg. Bicaudata Linn. Syft. Nat. No. 1. carries her eggs about with her, attached to the under fide of her abdomen, as fome fpiders are likewife known to do.

Schæffer has divided this Genus into the Perla and Phryganea, with the same distinct characters as Geoffroy; these two authors, I apprehend, were chiefly induced to pursue that method, because the number of joints, of which the Tarsi are composed, obliged them to arrange the different kinds of Phryganeæ under different orders.

The Antennæ of the Linnæan Phryganeæ, are filiform, and they have three stemmata.

The lesser Phryganeæ resemble the Tineæ so much, as not to be distinguished from them without difficulty; but, upon close examination, especially if the eye is aided by the microscope, the wings of the first are found to be almost covered with short hairs instead of the scales which adorn the wings of the Tineæ. The mouth of the Phryganea is likewise furnished with palpi, which are wanting in the Tineæ.

The larvæ belonging to this Genus, live in the water in tubes of silk, covered on the outside with small pieces of wood, sand, gravel, leaves of plants, &c. Nay, sometimes the larva attaches to its tube the smaller testaceous animals,

mals, yet alive, with their shells, and drags them about with it. They are much sought after by fishermen, by whom they are sometimes called *Stone*, or *Cod Bait*; the perfect insect is generally called the *Spring fly*, and is frequent near running waters, where the females resort to lay their eggs. They generally settle on the sides of walls, branches of trees, &c. which are least exposed to the sun, whose influence they seem to dread, seldom flying in the day time. Swallows are observed to feed upon them.

Genus IV. Hemerobius.

Linn. Syſt. Nat. page 911.

The mouth of the Hemerobius is armed with two teeth, and has four palpi.

The ſtemmata are wanting.

The wings are deflected, and not folded, as in the preceding Genus.

The antennæ are ſetaceous, advanced before the head, and longer than the thorax.

The thorax is of a convex form.

The Hemerobius is ſufficiently diſtinguiſhed from the Ephemera and Phryganea, by the poſition and formation of its mouth, which advances forwards, and is armed with teeth.

The Antennæ diſtinguiſh it from the following Genera belonging to this order.

Schæffer obſerves, that the abdomen grows ſlender towards its extremity, that the wings are in ſome ſubjects *incumbent*, in others *deflected*, and that the Tarsi are composed of five articulations.

Geoffroy has referred one species, the Hemerob. No. 12. of our author, to his *G. Phryganea*; and Linnæus himself appears doubtful to which of the two genera that insect belongs.

The under wings of most Hemerobii are of equal length with the upper ones; they are all four much weaker than in the preceding neuropterous genera, which makes their flight slow and unsteady.

Some of them are found near standing waters, others frequent gardens and fields; most of them are very ill scented. Their larvæ feed chiefly upon the aphides, of which they are exceedingly fond; but they sometimes devour other insects, and even one another.

One species belonging to this Genus, is known among us by the name of the *Golden Eye*.

Genus V. Myrmelion.

Linn. Syft. Nat. page 913.

The mouth of the Myrmelion is armed with jaws, two teeth, and four long palpi.

It has no ftemmata.

The tail, in the male fex, is furnifhed with a kind of forceps, formed by two ftraight filaments.

Their antennæ are club-formed, and as long as the thorax.

Their wings are deflected.

Geoffroy, who has defcribed only one fpecies of the Myrmelion, does not obferve that the tail of the male is furnifhed with a forceps. The fpecies he defcribed was perhaps incompleat, or he had met with none of the male fex; he adds that the four wings are all of equal length, and has given to that infect the generical name of *Formicaleo*, in which he is followed by Schæffer, who obferves that the wings in that genus

genus are deflected, and the tarsi composed of five articulations.

The last mentioned author has given the generical name of *Libelloides* to another species, in which the tail is forcipated, and the antennæ as long as the body, and the abdomen as broad as the thorax.

The Myrmeleon differs chiefly from the Hemerobius, under which Genus Linnæus had arranged it in the tenth edition of his *Systema Naturæ*, in the form of the antennæ, which are much shorter than those of the Hemerobius, in which Genus they are likewise setaceous: the male Hemerobius also wants the forceps which terminates the tail of the Myrmeleon.

The larva of the Myrmeleon lives chiefly upon ants; the perfect insect is very rare, but is sometimes met with in sandy places, and near rivulets.

Genus VI. Panorpa.

Linn. Syft. Nat. page 915.

The Panorpa has a horny, cylindrical probofcis, with two palpi.

It has two ftemmata.

The antennæ are longer than the thorax.

The tail in the male fex is furnifhed with a chela or weapon, refembling the claw of a crab, or the dart of a fcorpion.

The probofcis and tail of this infect render it too remarkable to be confounded with thofe of any other genus. The following characters, however, may be added to thofe of Linnæus, viz.

The wings extended horizontally on the back, when at reft.

The upper and under wings of equal length.

The palpi feated at the extremity of the probofcis.

The

The tarsi composed of five articulations.

The compleat insect is very common in the fields during the summer season, but the larva and chrysalis are yet unknown.

It has been called by some the *Scorpion-fly*.

Genus VII. Raphidia.

Linn. Syft. Nat. Pag. 915.

The head of the Raphidia is of a horny fubftance, and depreffed, or flattened.

The mouth is armed with two teeth, and furnifhed with four palpi.

The ftemmata are three in number.

The wings are deflected.

The antennæ are as long as the thorax, the anterior part of which is lengthened out, and of a cylindrical form.

The tail of the female is terminated by an appendix refembling a flexible, crooked briftle.

Schæffer obferves, that the antennæ of the Raphidiæ are fetaceous, and their tarfi compofed of four articulations.

According to Geoffroy, the wings are *incumbent*, rather than *deflected*, and the antennæ filiform.

The Raphidia is rarely to be met with; it is chiefly found in woods and hedges.

Linnæus fays, that the pupæ of one fpecies is (though it wants wings) exceedingly like the mother.

The larva has not been defcribed.

ORDER V.

INSECTA HYMENOPTERA.

The insects belonging to this order have generally four membranaceous naked wings; the Neuters, however, in some of the genera, and in others, the males or females want wings.

The tail (except in the male sex) is armed with a sting.

This order contains the following genera.

Genus I. Cynips.

Linn. Syst. Nat. page 917.

The mouth, in this genus, is armed with jaws, but has no proboscis.

The sting, which is spiral, is mostly concealed within the body.

Geoffroy

Geoffroy, who has confined the genus Cynips to such of the Linnæan species as have antennæ containing no more than thirteen articulations, and bent at their middle, or forming an angle, observes, that those insects have three stemmata; that their antennæ are cylindrical, and of equal thickness in their whole length; that their under wings are shorter than the upper ones; that their abdomen is nearly of an oval form, acute underneath, a little flattened on the sides, and attached to the thorax by a short stalk or pedicle, and that their sting is not placed at the extremity of their abdomen, but under that part, between two projecting plates, which form a kind of crest. This genus he has formed into three families; the first containing those species in which the antennæ are composed of eleven, the second those which have seven, the other those which have thirteen articulations.

He has arranged others of them, which have filiform antennæ not bent in their middle, and composed of fourteen articulations, under a new genus, which he terms *Diplolepis*; these, however, do not seem to differ generically from the Cynips, all the other characters assigned to them being the same as in that genus: the larvæ of the two genera likewise perfectly resemble one another, and live in the same manner

manner under the galls of plants, caufed by the infertion of the eggs by the females.

The genus termed by the fame author *Eulolophus*, of which he only defcribes one fpecies, feems, by his account of the larva, to be a Linnæan Cynips, with branched or pectinated antennæ; the fting, however, is extended from the extremity of the abdomen, and not from under that part. Linnæus has placed it in the laft family of his Ichneumons, *Vid. Linn. Syft. Nat. pag.* 941, *Ich.* 1, *No.* 77.

Schæffer, who has not feparated Linnæus's Cynipedes, obferves that their thorax is convex, and their wings extended, without folds, and that their tarfi have five articulations.

The gall made ufe of in the compofition of ink, is formed by an infect belonging to this genus.

Genus II. Tenthredo.

Linn. Syst. Nat. page 920.

The mouth of the Tenthredo is armed with jaws, but has no proboscis.

The wings are extended, and look as if swelled, or of a bulky consistence.

The sting, which is almost entirely hid within the abdomen, is dentated like a saw, and composed of two laminæ.

Two small tubercules are placed upon the scutellum at some distance from one another.

The antennæ of the Tenthredines, differing very much in their formation and number of their articulations, Linnæus has divided them into different families, taken from these circumstances, as follows:

1. Those with clubbed antennæ.

2. Those whose antennæ appear to be one continued thread, without articulations.

3. Those

3. Those with pectinated antennæ.

4. Those which have antennæ nearly clavated, or with a club less observable than that in the first family, and which are articulated.

5. Those with filiform antennæ, composed of seven or eight articulations, besides the base.

6. Those with setaceous antennæ, composed of several articulations.

Geoffroy and Schæffer have separated the first of these families from the others (though their larvæ and metamorphosis argue them not to differ generically) and have given their new genus the name of *Crabro*, with the following characters:

The antennæ club formed.

The under wings shorter than the upper ones.

The mouth armed with jaws.

The sting placed at the extremity of the abdomen, and serrated.

The abdomen throughout of equal fize, and clofely joined to the thorax.

Three ftemmata.

The remaining Tenthredines which, according to Geoffroy, have filiform antennæ, that author has divided into three families; the firft compofed of fuch as have nine; the fecond of fuch as have eleven, and the third of fuch as have thirteen articulations in their antennæ. The fame author obferves, that the under wings, likewife, of the *Tenthredo*, are fhorter than the upper ones; that the abdomen is clofely united to the thorax, not joined to it by a petiolum, or little ftalk, as in the Cynips, nor becoming fmaller from its extremity towards its bafe, fo as to form a kind of petiolum, as in the Ichneumon; and that the antennæ differ from thofe of the laft-mentioned genus in the form of their articulations, thefe in the Tenthredo are long, and rather rough, which makes their antennæ appear as if compofed of fo many knots; thofe of the Ichneumon, on the contrary, are fo very fhort as fcarce to be perceptible, and exceedingly fmooth, fo that, if not attentively examined, the antennæ would appear to be inarticulated, or like a briftle.

Scopoli

Scopoli, who has only described a small number of Tenthredines, has divided them into two families; the first containing those with elavated, the other, those having filiform antennæ, with seven or eight articulations: these last, he observes, turn aside, or bend downwards their antennæ, when under apprehensions of danger. The different sexes in this genus are in general differently coloured, which circumstance renders the knowledge of the species very difficult.

The larva of the Tenthredo differs entirely from that of all the other Hymenopterous insects, and resembles that of the Butterfly and Moth so much as easily to be mistaken for one of them: this resemblance has induced some Entomologists, who had attributed the term *caterpillar* to the larvæ of lepidopterous insects alone, to call those of the Tenthredo *false caterpillars*; there is nevertheless one certain rule to distinguish them by, that is, by examining the number of their feet; these, in the true caterpillar, never exceed sixteen, and are seldom so many; those of the false one, on the contrary, always exceed that number, being generally from eighteen to twenty-two: the six first, or pectoral ones, are hard, or scaly, and terminate each in a point, as those of the true caterpillar; the remaining

maining ones are soft and membranaceous, but deprived of the crotchets which terminate the membranaceous feet of the others: besides this distinction taken from the number of the feet, their heads are formed very differently; that of the false caterpillar consisting of one hard scale; that of the true one, on the contrary, is composed of two pieces, or scales, which Geoffroy calls *hoods*; the eyes of these last are likewise much larger than those of the others.

The larvæ of the Tenthredines feed chiefly upon the rose and willow trees, and undergo their last changement in the earth; their shrowd, or web, resembles net-work, being composed of large silken threads, between each of which great spaces are left, perhaps to let the humidity of the earth pierce to the chrysalis; the least excess of humidity or dryness in the earth kills those chrysalids, for which reason it is very difficult to bring them to perfection in boxes: out of more than three hundred larvæ of Tenthredines, which were nourished by Geoffroy, no tmore than five or six succeeded, though he took the utmost pains to keep the earth in a proper state. *Vid. Geoff. Paris. t.* 2, *p.* 269.

The Tenthredo is called, by some English Authors, the *Saw-fly*, from the formation of its sting, which differs from that of all other insects (that of the following genus only excepted) in being dentated or armed with teeth, like the instrument from which its name is taken; this sting, however, is not in the least dangerous, its weakness preventing the insect from doing any mischief with it.

Genus III. Sirex.

Linn. Syſt. Nat. page 928.

The mouth of the Sirex is armed with two ſtrong jaws.

The palpi, which are two in number, are truncated.

The antennæ are filiform, and contain upwards of twenty-four articulations.

The ſting is dentated like a ſaw, projected, ſtrong, and ſtiff.

The abdomen is ſlender, and terminates in a point or ſpine, from under which the ſting projects.

The wings are lanceolated (their extremities being drawn to a ſharp point) and are extended their whole length, not folded as thoſe of the Veſpa.

Scopoli has arranged the Sirices along with the Ichneumons, as Linnæus had likewiſe done in the former editions of his Syſtema Naturæ;

those insects, however, differ very much in their external appearance, formation, and manners; the abdomen of the Sirex is as broad as the thorax, and closely connected with, or joined to, that part: the abdomen of the Ichneumon, on the contrary, is either joined to the thorax by a petiolum or stalk, or grows much larger towards its extremity than at its base; the sting of the female Ichneumon terminates the abdomen, and is of a cylindrical form; that of the female Sirex projects from the under side of the abdomen, is dentated like a saw, and the abdomen itself is terminated by a kind of horn or spine. The female of the Ichneumon lays her eggs in the bodies of other insects (which she pierces for that purpose with her sting) and particularly in the bodies of caterpillars of Lepidopterous insects, upon which the larvæ feed, and where they remain till prepared for the chrysalis state; the female Sirex lays her eggs in the interior of decayed trees; the larva most probably feeds upon the wood, and always undergoes its last metamorphosis in the place where it had lived while in the caterpillar state: From all these circumstances, we may I presume safely conclude, that the Sirex differs generically from the Ichneumon.

Geoffroy

Geoffroy has only described one species belonging to this genus; to that insect he has given the generical name of *Urocerus*, a name taken from the point which terminates the abdomen, and which it were to be wished that Linnæus had adopted, since he himself looks upon the needless multiplication or changement of trivial names as a fault.

Schæffer has followed Geoffroy in the names and characters of this genus. These two authors add to the Linnæan characters, that the tarsi are composed of five articulations, and the under wings shorter than the upper ones.

The Sirex is very rare to be met with, but several species of it have been caught in England. It is generally called the *Tailed Wasp*.

Genus IV. Ichneumon.

Linn. Syst. Nat. page 930.

The mouth of the Ichneumon is armed with jaws, without any tongue.

The antennæ contain more than thirty articulations.

The abdomen is generally joined to the body by a pedicle, or stalk.

The sting is exserted, or projects beyond the abdomen, and is inclosed in a cylindrical sheath, composed of two valves.

The Ichneumons are divided into families, from the colour of their scutellum and antennæ, as follow:

1. Those with the scutellum white, and the antennæ noted with a white ring, or circle.

2. Those with a white scutellum, and black antennæ.

3. Those whose scutellum is of the same colour with the thorax, and which have a white ring on the antennæ.

4. Those with the scutellum of the same colour as the thorax, and the antennæ black and setaceous.

5. Those whose antennæ are yellow and setaceous.

6. Those with filiform antennæ, having the abdomen of an oval and slender form. The antennæ in this family often contain no more than ten articulations, the first of which is much longer than the others, and the insects in general are much smaller than the preceding ones.

Scopoli has united the Sirices of Linnæus with this genus, dividing it into two families; the first containing the last-mentioned insects; the second, the Ichneumons. He asserts, that the under wings, in the first family, are folded, the second he has subdivided from the colour of their antennæ.

Geoffroy adds to the Linnæan characters of the Ichneumon, that its antennæ are in a continual trembling motion; that the upper wings are longer than the under ones, and the stemmata three in number. That author has arranged some Ichneumons belonging to the last family, which have the abdomen of an oval form, under the first family of his Cynips.

Linnæus (as has been before observed) has placed the *Eulophus* of Geoffroy under this genus, from which that insect differs in its antennæ, which are pectinated.

Its larva, from the account given of it by Geoffroy, must resemble that of the Cynips; but Linnæus asserts that, like that of the Ichneumon, it lives in the bodies of other larvæ.

The species of Ichneumons are not easily determined, the different sexes varying much in their colours, nor can the distinct specific characters be well taken from any other circumstance.

Many

Many apterous insects are found, which, without doubt, belong to this genus; these very much resemble the apterous *Mutillæ*, from which they are distinguished, when living, by the continual vibration of their antennæ, which motion is not observed in the antennæ of the Mutillæ, and after death, by the roundness of their thorax, which is less retuse than that of the other genus, and by their long and slender abdomen, which is likewise frequently joined to the thorax by a petiolum. They are distinguished from the Sphex, which they likewise resemble, by the number of articulations in their antennæ.

Some of these apterous Ichneumons are, without doubt, females, having the sting, through which that sex deposit their eggs; others of them appear, from their being deprived of that sting, to be males. Geoffroy however asserts, that they are all females: perhaps that author had only met with such as had stings.

The larvæ of many Ichneumons not only live, but likewise undergo their metamorphosis, in the chrysalids or caterpillars of Lepidopterous insects; others of them, when arrived at their full growth, pierce the skins of their lodgments,

ments, which they quit, and fixing themselves to the sides of walls, branches of trees, &c. there pass the chrysalis state under cover of a silken web.

The name of *Ichneumon-fly* has been given to this genus, by some English authors.

Genus V. Sphex.

Linn. Syst. Nat. page 941.

The mouth of the Sphex is armed with jaws, but has no tongue.

The articulations of their antennæ are ten in number.

The wings, in each sex, are extended, without folds, and laid horizontally upon the back.

The sting, which is sharp and pointed, is concealed within the abdomen.

This genus is divided into two families; in the first of which, the abdomen is petiolated, or joined to the thorax by a stalk; in the other, the abdomen is subsesile, or of a slender make, nearly of equal size in its whole length, and attached to the abdomen without a petiolum.

Scopoli has divided his Spheges (to which he gives the same characters as above, excepting what relates to their antennæ) into three families;

the

the two first of which are Linnæan Spheges, and distinguished from one another by the same circumstances as by Linnæus; the third (the abdomen of which he says is sessile) contains the Chryses of our author, which differ from the Spheges in the formation of their antennæ, in the lateral scale of the abdomen, which the last-mentioned insects want, and in the spines which terminate the thorax and belly.

Geoffroy has placed such Spheges as were known to him among his Ichneumons, as Linnæus had likewise done, in the tenth edition of his Syst. Naturæ. It has already been shewn, that they differ from that genus in the number of articulations which compose their antennæ, and in the position of their sting, which in the last-mentioned genus is exserted.

Schæffer has, like Linnæus, separated the Spheges from the Ichneumons, and assigns them the following characters:

The tarsus of each foot composed of five articulations.

The antennæ club-formed, and bent.

The mouth armed with jaws, and furnished with palpi.

The

The ftemmata three in number.

The wings extended, incumbent, without folds, and the under ones fhorter than the upper ones.

The abdomen of an oblong form.

The fting pointed, and concealed within the abdomen.

A great number of exotic infects have lately been brought from different countries, which would certainly belong to this genus, if they were not provided with long membranaceous tongues, like thofe of the Bee, from which genus other circumftances again feparate them. Whether or no thefe infects differ generically from the Sphex does not appear to have been determined.

Many fpecies of this genus are common in England; they are chiefly found in woods and hedges; their larvæ feed upon dead infects, in the bodies of which they are produced from the egg; fome fpecies dig holes in the earth with their fore feet, like dogs, in which holes they bury dead infects, chiefly fpiders or Lepidopterous larvæ, and after having depofited their eggs

in

in the bodies of these insects, they carefully close the holes with earth.

It is very probable that some species of Apterous Spheges are found in England, which matter must be determined by the external appearance, the sting's being concealed within the abdomen, and the number of articulations in the antennæ.

The Sphex is called by some, the *Ichneumon-Wasp*.

Genus VI. Chrysis.

Linn. Syst. Nat. page 947.

The mouth of the Chrysis is armed with jaws, but has no proboscis.

The antennæ are filiform: the first articulation is long in proportion to the exterior ones, which are eleven in number.

The abdomen is elevated in the middle, like an arch (*fornicatum*) with a kind of lateral scale on the under side.

The anus is dentated, or terminated by teeth or spines, and likewise armed with a sting, which projects a little.

The wings are extended, not folded, as in the Vespa.

The body is of a shining colour, and appears as if gilt.

Scopoli, as before observed, has arranged the Chryses among his Spheges.

Geoffroy

Geoffroy has placed them among the Vespæ, as Linnæus had done in the tenth edition of his System: he has, however, formed them into a separate family, under the title of *Golden Wasps*. They differ chiefly from that genus in the position of their wings, which are not folded, and in the spines situate on each extremity of the thorax, in most species of the Chrysis.

Schæffer has adopted Linnæus's method in preference to that of Geoffroy, adding to the characters of that author, that the antennæ are bent and cylindrical; that the tarsus of each foot is composed of five articulations; that the four wings are all equally transparent, and have very few nerves or membranes; and that the abdomen is oval, and of equal size with the thorax.

The Chrysis lives chiefly in the holes of old walls, where they likewise lay their eggs: their larvæ resemble that of the Wasp.

GENUS VII. VESPA, the WASP.

LINN. Syſt. Nat. Pag. 984.

The mouth of the Veſpa is armed with jaws, but has no tongue.

The upper wings are folded in both ſexes.

The ſting, which is ſharp and pointed, is concealed within the abdomen.

The body is ſmooth, without hair.

The eyes (as obſerved by De Geer) are lunular.

Geoffroy aſſigns the following characters to the Waſp.

> The antennæ bent, with the firſt articulation very long in proportion to the others.
>
> The inferior wings ſhorter than the upper ones.
>
> The mouth armed with jaws, and provided with an *inflected membranaceous tongue*.
>
> The ſting ſmooth and pointed.

The abdomen attached to the thorax by a short pedicle.

Three stemmata.

The body smooth, without any hairs upon it.

From the above it will appear, that Linnæus and Geoffroy differ very essentially with regard to one character assigned by the latter to the Wasp, viz. that of its having a membranaceous tongue, the existence of which Linnæus denies, but which, according to the other, is placed in the mouth between the jaws, bent inwards under the breast, and composed of several pieces or membranaceous filaments, exactly like that of the Bee; this difference, in a matter to all appearance so easy to be decided, is surprising: No author, besides Geoffroy, that I am acquainted with, pretends that the Wasp has a tongue, nor could I ever perceive it, though I have purposely examined a great number of European Wasps, and particularly such species as are described by that author, and which were taken in France; all, indeed, have a kind of broad, membranaceous skin under the jaws, at the base, or upon the sides of which, the palpi are seated; this membrane does not, however, in the least resemble a tongue, nor does it seem calculated to

serve

serve instead of one; it has the appearance of a little bag with the mouth downwards, but does not close on the under side; towards the end it is jagged, and divided into lobes, exactly like the petals of some flowers. If Geoffroy took this membrane (which is always very short) for a tongue resembling that of the Bee, he was certainly mistaken, or had not examined it with sufficient attention.

Linnæus's character perhaps ought not to be taken for generical, since he himself describes one exotic species, and several others are found in the cabinets of the curious, which are provided with tongues; these, indeed, differ very much from the tongue of the Bee, being (in such species as I have met with, and particularly in two or three which I possess myself) short, stiff, extended, and concealed under the upper lip, which for that purpose is drawn or lengthened out into a horny, pointed proboscis; the bodies of some of these insects are hairy, like Bees, others are smooth, or without hairs. It is to be hoped that some ingenious traveller will take upon himself the task of examining whether or no these last-mentioned insects differ generically from the Wasp and the Bee, or to which of them the different species belong, which can only be done by those who shall have opportunities of examining their manner of living and metamorphosis.

Schæffer says, that the mouth of the wasp is furnished with palpi, but does not mention the tongue; the tarsi, according to him, are composed of five articulations.

Scopoli says, that the wasp has no tongue.

Many kinds of Wasps live in societies, after the manner of Bees, and like them make combs, in which they deposit their eggs; they likewise feed their larvæ with honey, but of a very inferior quality to that of the Bee; others of them construct a different or separate nest for each egg.

The larvæ and chrysalids of all of them resemble that of the Bee.

GENUS

Genus VIII. Apis, the Bee.

Linn. Syſt. Nat. page 953.

The mouth of the Bee is armed with jaws, and furniſhed with a proboſcis, incloſed in a bivalve ſheath, and inclined downwards under the body.

The wings are extended, and without folds in each ſex.

The females and neuters carry a ſharp pointed ſting concealed in their abdomen.

This Genus is divided into two families, the firſt containing ſuch as have the body ſmooth, without any, or with very few hairs; the ſecond, compoſed of thoſe whoſe bodies are very hairy, and which emit a found as they fly.

Scopoli having obſerved that the quantity of hair on the bodies of the different ſpecies of Bees, encreaſes ſo gradually, as likewiſe the

noiſe

noise they make in their flight, as to render it difficult to determine where the first family of Linnæus shall end, or the other commence, has therefore preferred to divide them into families, from the form of their antennæ, which in some are whole and extended, in others bent, and forming an angle from their base; this division seems liable to fewer inconveniencies than that of Linnæus, though it frequently connects Bees which differ much in their outward appearance.

Geoffroy observes that the under wings of the Bees are shorter than the upper ones; that the first articulation of their antennæ in each sex, is much longer than the others; that the abdomen is join'd to the thorax by a short pedicle, and that they have three stemmata. He has divided them into families for the same circumstances as Linnæus.

The tarsi in this Genus are composed of four articulations.

The Bee is too well known to be easily confounded with any other Genus of Insects. The female of the domestic Bee is much larger than the male or neuter; her antennæ contain fifteen articulations; her abdomen is composed of seven segments, and is much longer than her wings. The antennæ of the male contain only eleven articulations, nor has that fex any fting; the neuters are much fmaller than the males or females, their antennæ contain fifteen articulations; they are likewife remarkable by the hairinefs of the under fide of their hindmoft thighs, which refemble a kind of brufh, with which they gather the fine powder fcattered from the *Antheræ* of flowers, and from which the wax or comb is made.

The induftry of thefe little animals, which is as profitable as curious in itfelf, will always continue to excite the admiration of the wifer part of mankind. Swammerdam, Reaumur, Hagftrom, D'Aubenton, Geoffroy, and other authors, have wrote their hiftory with great accuracy. Swammerdam, above all, deferves to be read with the greateft attention.

Genus IX. Formica, the Ant.

Linn. Syſt. Nat. page 962.

The Formica (called among us the Piſmire, Emmit, or Ant) is diſtinguiſhed by the little upright ſcale which is ſituate between the thorax and the abdomen.

The ſting with which the females and neuters are armed, is concealed within the abdomen.

The males and females are winged, the neuters apterous.

To the above characters of the Ant, Geoffroy adds that the antennæ form an angle, their firſt articulation being very long in proportion to the others, that the ſtemmata are three in number, and the abdomen joined to the thorax by a ſhort ſtalk.

Schæffer likewiſe adds, that the mouth is armed with jaws, that the wings are incumbent,

bent, and the tarfi compofed of five articulations.

The Ants live in focieties compofed of males, females, and neuters; the males are much fmaller than the females and neuters, but are diftinguifhable from the largenefs of their eyes, which are not fo well proportioned to the fize of their body as in the other fexes.

No fooner is the work of generation performed, than the male and female ants perifh, as well as moft of the neuters; fome of thefe, however, outlive the winter, but pafs that feafon in their habitation, without movement, or any figns of life. How ufelefs then would be that prudence and affiduity in laying up a ftock of provifions for the winter, attributed, for fo many ages, to the Ant?

The female Ant feems to take no farther care of the young, after having depofited her eggs; the important office of nourifhing the larvæ, and preferving the chryfalids, is entirely left to the neuters, whofe affection for a progeny neither begot nor brought forth by them, can never be fufficiently wondered at; they labour inceffantly to fupply the larvæ with provifions,
and

and are constantly employed in preserving the chrysalids from humidity in wet seasons, or exposing them to the warmth of the sun when it is fair. These chrysalids are much larger than themselves, yet are carried about by them with ease; many kinds of birds are very fond of, and devour them, as well as the Ants themselves.

GENUS

Genus X. Mutilla.

Linn. Syst. Nat. page 966.

The Mutillæ, for the most part, want wings.

Their body is covered with a kind of down.

The thorax strikes off bluntly at its base, or rises perpendicularly from the part where joined to the abdomen.

The sting is pointed, and concealed within the body.

The Mutillæ are as yet very little known, only two or three species have been found with wings, and we are ignorant whether these are *males* or *females*; perhaps they live in society like the Ant, and the apterous species are neuters. Most of the insects without wings, arranged by different authors under this Genus, appear to be either Ichneumon's or Spheges; that described by Scopoli was most probably an Ichneumon, from the vibrating motion of its antennæ; and

Linnæus

Linnæus himself is of opinion, that two of the five European species described by him, belong rather to the last-mentioned Genus than to the Mutilla; these two species, as well as the *Mutilla Europæa, Linn. No.* 4. have been found in England, but their manner of living, their larvæ, and metamorphosis, are wholly unknown, as the Genus itself appears to have been to Geoffroy and Schæffer, since neither of these authors has described any of the species belonging to it.

ORDER VI.

INSECTA DIPTERA.

The infects belonging to this order have two wings.

They are furnished with a poiser or balancer, (Halteres) situate under each wing, which is terminated by a capitulum or knob. The base is concealed or secured under a little scale, by which it is covered as by a shed.

This order contains the following genera, viz.

GENUS I. OESTRUS.

LINN. Syst. Nat. Pag. 969.

The Oestrus has no mouth, in the place of which three small impressed points are found

found, without any visible proboscis or rostrum.

Geoffroy observes, that the antennæ of the Oestrus are setaceous, and grow, or are placed, upon a small point or button.

They have three stemmata.

Frisch, in his description of the Oestrus Bovis, Linn. No. 1, asserts, that that insect has a rostrum, which it can draw within its head, and shoot out at pleasure.

Schæffer observes, that the abdomen in this Genus is of equal size with the thorax.

The larvæ of the Oestri lay hid in the bodies of cattle, where they are nourished the whole winter; the perfect insects are to be met with in the summer almost wherever horses, cows, or sheep are grazing; some of them lay their eggs under the skin of cows or oxen, which they pierce for that purpose; others, for the same end, enter the intestines of horses by the anus, and others, again, deposit them in the nostrils of sheep; in these different habitations

the

the larvæ refide till full grown, when they let themfelves fall to the earth, and generally pafs the chryfalid ftate under cover of the firft ftone they meet with.

The Oeftrus is in fome places known by the name of the *Gad-Fly*.

Genus II. Tipula.

Linn. Syst. Nat. page 970.

The head of the Tipula is long, or seems lengthened out.

The upper jaw is formed like an arch.

The palpi are two in number, curve, and longer than the head.

The proboscis is short, and bent inwards.

They are divided into two sections, the first containing those in which the wings are open or extended when at rest; the other those whose wings cover the body horizontally when sitting.

Scopoli has divided the tipulæ into two families, in the first of which their antennæ are pectinated in the males, both sexes in the other have simple antennæ.

Geoffroy

Geoffroy has selected some of the Linnæan Tipulæ, in which the antennæ appear to be perfoliated, and are shorter than the head, and arranged them under a new genus, to which he has given the name of BIBIO. The antennæ, in the figure he has given of the Bibio, appear rather to be formed of large articulations, growing regularly smaller towards the extremity, than perfoliated; In the antennæ of the Bibio, figured by Schæffer, the articulations seem to be largest in the midfile, and to decrease in size towards the base and extremity. The larvæ of the Bibiones differ very essentially from those of the Tipulæ, in the number of their stigmates, which, like those of the caterpillars of Lepidopterous insects, are arranged along the body on each segment; in the Tipulæ, they are but four in number, two at the head and two at the tail; these last are found in the trunks of decayed trees, those of the Bibio are most frequent in the dung of cows. The Tipulæ have three stemmata.

They are called, by some English authors, *Crane-Flies*: many kinds of fish, birds, and larvæ of aquatic insects devour them.

Genus III. Musca.

Linn. Syft. Nat. page 979.

The mouth of the Musca is formed by a soft, fleshy proboscis, with two lateral lips; it wants palpi.

The Muscæ are divided into different families, from the form of their antennæ, as follows:

1. *Filatæ*, with simple antennæ, or whose antennæ are without any lateral hair or feather.

2. *Armatæ*, in which the antennæ are furnished with a lateral hair, or feather; these last are either

    Tomentosæ, or *Pilosæ*.

    The bodies of the *Tomentosæ* are downy, though scarce perceptibly so; and they are either

Plumatæ

Plumatæ, having a lateral plume, or feather on the antennæ: Or

Setariæ, with a simple hair on the side of the antennæ.

The *Pilosæ* have a small number of hairs scattered upon their bodies, principally upon the thorax; they are either

Plumatæ, with a lateral feather: Or

Setariæ, with a lateral hair.

Geoffroy has divided the Linnæan Muscæ into the following genera:

1. *Stratiomys*: This genus comprehends such of them as have the hinder part of the thorax armed with spines, and the antennæ without any lateral hair or feather, and forming an angle from the end of the first joint, which is much longer than the others; it is farther divided into two families, the first having two, the other six spines, on the thorax.

The larvæ of this genus live in the water, and devour small aquatic insects; the fly itself is found frequently near pools of water, whither it resorts to lay its eggs.

2. *Musca*, composed of such Linnæan Muscæ as have solid antennæ, of a flattish form, somewhat resembling the mouth of a spoon in shape, and accompanied by a lateral hair; this genus he has divided into families, from the following circumstances:

1. Those whose wings are of various colours.

2. Those which have, on the fore part of the head, a kind of pelicle, or membrane, which appears as if swelled, and forms to the insect a kind of mask, generally of a light colour.

3. Those whose bodies are of various colours.

4. Those of a gold colour.

5. Those of the most common colours, or such as have nothing remarkable about them.

The larvæ of some of this genus devour the the Aphides; these larvæ seem to want eyes, and lengthen or stretch out their head as if to feel for their prey; others live in and consume all kinds of putrid flesh; others are found in cheese; others, again, in the excrement of different animals; many live in the water, and prefer the most putrid and muddy.

3. VOLUCELLA, which genus contains the *Muscæ Plumatæ* of Linnæus, or those whose antennæ are furnished with a lateral feather.

The mouth of this genus, according to Geoffroy, is formed by a proboscis concealed within a sheath.

The larva of the voluncella perfectly resembles that of the Musca, and is frequent upon the rose.

4. NEMOTELUS: this genus is composed of such Linnæan Muscæ as have moniliform antennæ ending in a kind of sharp point.

The mouth resembles that of the Voluncella.

5. SCATHOPSE, which differs from his Musca only in the shape of the antennæ, which are filiform.

Schæffer has adopted all these new genera, and observes that they have each three stemmata.

Scopoli has formed the following new genera from the Linnæan Muscæ, on account of the different formation of their proboscis, or antennæ:

1. *Musca*, to which he gives the following characters:

The mouth armed with a retractable proboscis, which is dilated at its extremity, and furnished with clavated palpi, situate at its base.

2. *Ceria*: the rostrum of the Ceria is formed like that of the Musca.

The antennæ are moniliform, with the last articulation larger than the others.

This genus belongs to the first family of Linnæan Muscæ.

3. *Conops*:

3. *Conops:* which genus is distinguished by the following characters:

The mouth armed with a quadrifeted rostrum, two of which setæ are longer than the others; the sheath of the rostrum is retractible, fleshy, and terminated by lips: the upper lip is formed by two lobes, the under one is bifid.

The setæ or bristles above-mentioned, are situate, in this and the following genera, at the base, and extended longitudinally towards the extremity of the rostrum.

The Conops is formed in part from such Linnæan Muscæ as have a lateral feather, in part from such as have a lateral hair on their antennæ; and Scopoli has divided them, from that circumstance, into two families.

4. *Anthrax:* the mouth of the Anthrax is armed with a bifeted rostrum; the sheath is fleshy at its base, and retractible; its extremity is simply dilated, not divided into lips, as in the Conops.

The palpi are seated in the middle of the rostrum.

Scopoli has only described one species of this genus, which is the *Musca Morio*, No. 9, of Linnæus.

The Muscæ are the most common of all insects, and are known to every one. The name of *Fly* is particularly applied to them.

Genus IV. Tabanus.

Linn. Syst. Nat. page 999.

The mouth of the Tabanus is extended into a fleshy proboscis, terminated by two lips.

The rostrum is furnished with two pointed palpi placed on each side of, and parallel to, the proboscis.

Scopoli assigns the following characters to the Tabanus:

The mouth armed with a proboscis, on which are five bristles; these bristles are seated (as in his *Conops*, *Anthrax*, &c.) at the base of the rostrum, and extended almost to its extremity.

The sheath is univalve and obtuse.

The palpi are two in number, acuminated, porrected, parallel, and incumbent upon the rostrum, so as to form a kind of second or upper valve to the sheath.

The specific characters of the Tabani are chiefly taken from the colour of the eyes, which

this

this author obferves ought to be examined while the infect is yet alive.

Geoffroy afferts, that the roftrum of the Tabanus is accompanied by two ftrong teeth, with which the infect pierces the fkins of horfes, &c. No other author has mentioned the exiftence of thefe teeth, nor could I ever perceive them.

The antennæ, according to the fame author, are of a conic form, and divided into four parts, being generally compofed of feven articulations, the three firft of which, from the bafe, are much larger than the four others, and form, as it were, three diftinct pieces; the four others are much fhorter, and appear as if confounded together, or forming only one piece; the third piece is generally larger than the two firft, and attended with a kind of lateral appendix, which makes the antennæ appear as if forked.

Schæffer obferves, that the Tabani have three ftemmata, and that their abdomen is as broad as their thorax.

The Tabani nourifh themfelves with the blood of horfes and cattle. As they are moft frequent
near

near watry places, it is probable that their larvæ are aquatic, though De Geer asserts that they live under the earth.

They have been named *Burrel* or *Whame Flies*, by some English authors.

Genus V. Culex.

Linn. Syst. Nat. page 1001.

The mouth of the Culex is formed by a flexible sheath, enclosing setæ, or bristles, pointed like stings.

According to Scopoli, the bristles of the rostrum in this genus are four in number, and two of them are longer than the others; the sheath is long and porrected, and the palpi are incumbent upon the base of the rostrum.

The antennæ of the female Culices are filiform, those of the males feathered. The thorax, in both sexes, is gibbous, and the abdomen attenuated, growing smaller from its base to its extremity; this part, in the females, is generally longer than the wings; in the male, on the contrary, it is much shorter: the wings, in both sexes, are extended horizontally along the abdomen. The Culices have no stemmata; they very much resemble the smaller Tipulæ, from which, however, as Geoffroy observes, they may be easily distinguished by their mouth, which, on comparing the characters given to the two genera, or the insects themselves, will appear to be formed very differently.

The

The larvæ of the Culices are very frequent in ſtanding waters; their bodies are compoſed of nine ſegments, which diminiſh in ſize and length from the head towards the extremity of the body; the laſt of theſe ſections is furniſhed with a kind of ſtigmate, through which the larva breathes, frequently riſing, for that purpoſe, to the top of the water. The head of the chryſalis is ſo much bent under the breaſt, that the thorax appears to be the moſt advanced part of the body; the ſtigmates are placed upon the back of the thorax; the ſegments of the abdomen diminiſh in ſize towards its extremity, the laſt terminates in a kind of flat tail or fin, by means of which the inſect ſwims or moves itſelf in the water.

The Culices generally frequent woods and watry places; they are known by the name of *Midges*.

Scopoli informs us, that where large quantities of them are found, the ſoil is generally marſhy, and the air unwholſome.

The females are very troubleſome, and ſting ſeverely, which the males are ſeldom obſerved to do.

GENUS

Genus VI. Empis.

Linn. Syst. Nat. Pag. 1003.

The proboscis of the Empis is of a strong horny substance, it is bivalve, inclined downwards under the head and breast, and longer than the thorax: the valves are horizontal.

Scopoli has placed the only species of this genus, which he has described among his *Asili*, to which genus he gives the following characters:

The mouth armed with a quadrifeted proboscis: the sheath porrected, stiff, longer than the head, and bivalve. He adds, that the head is small, of a roundish form, the back gibbous, the feet long, and the rostrum small and inflected.

According to Schæffer, the antennæ in this genus are composed of three articulations, the first of which is long and filiform; the second very short and globular; the third much larger at its base than in the middle, from whence, again, it grows larger, and is finally terminated by a long and sharp point.

The wings in this genus are incumbent.

Schæffer says that the antennæ are of a conic form.

The Empis seems not to have been known to Geoffroy.

The perfect insects are common upon flowers, and in gardens, but I do not find that the larvæ or chrysalids have been described by any author.

GENUS VII. CONOPS.

LINN. Syſt. Nat. page 1004.

The proboſcis of the Conops is porrected and jointed.

Scopoli has given the following definition of the genus, named by him *Empis*, and under which he has arranged ſome of the Linnæan Conopſides:

The mouth armed with an uniſeted proboſcis, which is membranaceous at its baſe (where the palpi are ſituate) and capable of being drawn in and extended; towards the end it is ſtiff, long, porrected, and attenuated.

The Stomoxys of Geoffroy is a Linnæan Conops, and the Empis of Scopoli; he deſcribes it as follows:

The antennæ terminated (like thoſe of the Muſcæ) by a flat and ſolid articulation, ſhaped like the mouth of a ſpoon, with a lateral briſtle, which, when cloſely examined, appears to be very hairy.

The

The mouth formed by a proboscis, which is shaped like an awl, simple and acute.

Three stemmata.

This insect very much resembles the *Musca Domestica*, or common Fly, but is distinguished by the different formation of its rostrum.

The genus termed *Sicus*, by Scopoli, contains two species of Linnæan Conopsides, viz. the Conops *Testacea*, No. 11, and the Conops *Buccata*, No. 12. The Sicus is distinguished by the following characters:

The mouth armed with an uniseted proboscis, with a stiff, porrected, and long sheath, broken or bent in the middle, and inflected.

The palpi seated at the base of the sheath.

The Sicus differs chiefly from the *Empis* of the same author in the formation of its proboscis.

The *Stomoxoides* of Schæffer is the *Sicus* of Scopoli; he has described it as follows:

The antennæ shaped like those of the Linnæan Muscæ, with a lateral hair.

The mouth formed by a porrected proboscis which is bent, or shuts like a clasp-knife.

Three stemmata.

The abdomen for the most part curve.

The *Rhingia* of Scopoli is likewise a Linnæan Conops; he describes it as follows:

The mouth armed with a trifeted proboscis; the middle bristle longer than the others, and bifid, the lateral ones (on which the palpi are seated) of equal length with one another; the sheath of the rostrum is univalve, attenuated, and applied to the canal of the mouth.

The Conops is chiefly found in meadows and fields, where the different species are very troublesome to cattle.

I do not know that the larvæ or chrysalids have been described.

Genus VIII. Asilus.

Linn. Syst. Nat. page 1006.

The rostrum of the Asilus is hard, or horny, porrected, extended out its whole length, and bivalve.

Scopoli has arranged many of the Linnæan Asili under the genus called by him *Erax*, to which he assigns the following characters:

The mouth armed with a trifeted proboscis, or on the base of which are seated three bristles, two of which are shorter than the others, on which the palpi are often seated.

The sheath, which does not exceed the head in length, is composed only of one valve.

The *Asilus* of that author differs from his *Erax* chiefly in the form of its proboscis, which contains four setæ, or bristles; the sheath is porrected, stiff, longer than the head, and bivalve.

Schæffer describes the Asilus as follows:

The antennæ with a bristle arising from a cone.

Three stemmata.

The mouth, with a proboscis, which is extended, horny, setaceous, and bivalve.

The thorax gibbous.

The abdomen attenuated.

The feet made for running.

The halteres very large.

The feet of the Asili, as Geoffroy observes, are large, and the articulations, which are five in number, short, and shaped like a heart.

The Asilus is called, by some authors, the *Wasp-Fly*, and not improperly, since, like the Wasp, it stings severely whatever offends it, though with a different instrument, viz. its
proboscis,

probofcis, for which reafon it ought not to be taken without precaution.

Many fpecies of them are not uncommon in watry meadows, where they very much incommode the horfes and cattle.

Its larvæ, or chryfalis have not been defcribed that I know of.

Genus IX. Bombylius.

Linn. Syst. Nat. page 1009.

The rostrum of the Bombylius is porrected, setaceous, very long, and formed by two horizontal valves, in which are contained setaceous stings or bristles.

Scopoli, who describes under this title only the same species found in the *Systema Naturæ*, observes, that the proboscis is long, porrected, and bivalve, and that the upper valve is entire at its extremity, bearded, and shorter than the under one, which last is bifid at its end, and not hairy; that the *two* palpi are depressed, and seated at the base of the inferior valve; and that the bristles at the base of the proboscis are two in number.

Geoffroy has placed the only one of this genus which he had met with, among his Asili, from which genus it differs in the number of bristles seated at the base of the proboscis, which are four in number in the Asilus; in the length of the proboscis, which part is much longer in the Bombylius than that of the Asilus; and in the position of the wings, which in the last-mentioned

tioned genus are crossed one over the other, but in the other are open.

Schæffer observes, that the antennæ are broken or bent, setaceous, and of a conic form; that the stemmata are three in number; that the abdomen is as broad as the thorax, and the wings patent, or open.

Several species of the Bombylii are very common in the spring about the months of March and April; they are generally found upon flowers in woods and low marshy grounds.

Their larvæ are probably aquatic, since the perfect insects frequent waters. I do not know that they have been described.

Genus X. Hippobosca.

Linn. Syst. Nat. page 1010.

The rostrum of the Hippobosca is bivalve, cylindrical, obtuse, and wavering or shaking, as if ill fixed to the mouth.

The feet are armed with many nails, or crotchets.

Scopoli adds to these characters, that the rostrum has only one bristle.

Geoffroy observes, that the Hippoboscæ are remarkable in being the only genus of Dipterous insects which want stemmata, except only the *Culex*; their antennæ are setaceous, very short, and composed of a single hair; they are very flat, hard, and as it were scaly: it is very difficult to kill them by compression.

The wings, in some subjects, are crossed one over the other, in others are open.

Schæffer observes, that their abdomen is as broad as the thorax.

The Hippoboscæ have been called, by some authors, *Spider-Flies*, from the great resemblance

blance which one of them bears to that insect; others have called them *Horse-Flies*, by which name they are more generally known; they are found frequently in woods and marshy places, but most commonly on the bodies of birds, horses and other quadrupedes, sucking their blood, upon which alone they subsist. Their larvæ are unknown. One of the species is known to be pupiparous; the egg of this insect is larger than the mother, and is rather a pupa or chrysalis, than a real egg, since the compleat or winged insect is produced from it.

ORDER VII.

INSECTA APTERA.

This order contains all such insects as want wings in either sex.

It has been before observed, that many insects are found to want wings, which, however, cannot be referred to this order, because one or other sex of the same species is furnished with those parts. *Bruniche*, in his System of Entomology, has indeed arranged every insect wanting wings under his *Apterous order*, without taking notice of the wings in the different sexes of the same species, which creates a strange confusion, as the different sexes of the same insect must often be sought for under different orders: thus the Apterous Aphis, the female Coccus, the neuters of Ants, the Apterous Mutillæ, are separated from the others of their own species, and arranged among insects with which they have no affinity. He has likewise placed the pupa of the Gryllus under this order, which is doubly improper for the reason above mentioned, and as not being a compleat insect.

The whole species, or every sex of the same insect, must want wings in order to render it *apterous* in the sense of Linnæus, and to place it under this order, which contains the following genera:

Genus I. Lepisma.

Linn. Syst. Nat. page 1012.

The Lepisma has six feet, formed for running.

The mouth is furnished with four palpi, of which two are setaceous, and two capitated.

The tail is terminated by extended bristles.

The body is imbricated with scales.

Schæffer asserts, that the Lepisma has only two palpi; that its antennæ are setaceous; the briftles of its tail *three*, and its eyes *two* in number; and that the fix feet are broad and fcaly at their bafe, and formed for running.

Geoffroy, who has given the name of *Forbicina* to this genus, fays, likewife, that the feet are broad and fcaly at their bafe. He is of opinion, that the antennæ are fetaceous rather than filiform.

Scopoli obferves, that the tail of the Lepifma is not made for leaping as that of the *Podura*.

The infects belonging to this genus are very frequent under old floors, wainfcots, &c. efpecially in damp houfes; they run with great fwiftnefs, and are generally of bright, fhining colours; they are fuppofed to live upon Wood-Lice, or by fucking the humidity of the wood under which they live.

Genus II. Podura.

Linn. Syst. Nat. page 1013.

The Podura has six feet, which are formed for running.

The eyes are two in number.

The tail is forked, bent inwards under the body, elastic, and acts like a spring, by which the insect leaps.

The antennæ are long and setaceous.

Schæffer says, with Geoffroy, that the body is covered with scales, and the last-mentioned author has divided the Poduræ into two families; the first containing those of a short and globular form; the other, those of a long and slender make; the antennæ, according to the same author, are filiform.

The Podura pretty much resembles the *Pediculus*, from which it differs principally in its tail; that part, when the insect is at rest, or walks undisturbed, is bent under the abdomen, and preserved in a kind of groove, from which, when

when inclined to leap, the infect withdraws it, and by ftriking it with force againft the ground, is thrown to a confiderable diftance.

The Poduræ are generally found upon the ground in fand or gravel-pits, or under branches of trees, ftones, &c. in humid places. One fpecies is found upon the water, upon the furface of which it leaps with great agility. It is not known upon what any of them feed.

Genus III. Termes.

Linn. Syst. Nat. page 1015.

The Termes has six feet made for running.

Two eyes.

Setaceous antennæ : And

The mouth armed with two jaws.

Scopoli says, that the Termes resembles the Pediculus, or Louse, and Geoffroy has described the only one he knew as such. *Vid. Geoff. Paris.* 2, *p.* 601, *ped. No.* 12.

They are generally called *Wood-Lice*.

GENUS IV. PEDICULUS.

LINN. Syst. Nat. page 1016.

The Pediculus has six feet formed for walking.

It has two eyes.

Its mouth contains an exserted sting.

The antennæ are as long as the thorax.

The abdomen is depressed, and as it were formed of different lobes.

But few of the pediculi of quadrupedes and birds have been observed, and the specific characters of still fewer determined; though it is pretty certain that almost every different animal is infested with a different species of them.

Schæffer says, that the antennæ of the Pediculus are setaceous, and the head distinct from the thorax, which parts appear to him to be united in some other genera belonging to this this order.

The Pediculi are of various forms or shapes; some of them are almost oval, others oblong, other again very long and slender; their head is large, their eyes prominent, and their abdomen composed, in some, of more, in others, fewer segments, from six to ten; their tarsi are composed of three articulations; the crotchet, or nail, is semilunular, and very sharp.

They are oviparous animals, and their eggs are pretty large; they change their skin several times before they are full grown; they are thought to be hermaphrodites, which circumstance may account in part for their prodigious multiplication.

Swammerdam, who had dissected a great number, and has given a very good history of them, assures, that he never found one without an ovary, nor ever found the exterior parts of generation peculiar to the male sex. If they are all formed thus, the Louse is an hermaphrodite of a very particular kind, and must be able to fœcundate itself without copulation, which no other animal can do. Many kinds of *vermes*, or *worms*, are hermaphrodites, but far from being able to fœcundate themselves, they have occasion for a double copulation,

copulation, each individual performing the office both of male and female. This matter deserves the serious attention of Entomologists, and may be determined, perhaps, without great difficulty, these insects being so common. *Vid. Geoff. Paris : tom.* 2, *pag.* 506.

Genus V. Pulex, the Flea.

Linn. Syſt. Nat. page 1021.

The feet of the Flea are ſix in number, and formed for leaping.

It has two eyes.

The antennæ are filiform.

The roſtrum is bent inwards, ſetaceous, and conceals a ſting.

The abdomen is compreſſed or flattened.

Schæffer obſerves, that the body of the Flea is covered with ſcales.

The roſtrum, according to Scopoli, is bivalve.

The Flea is the only inſect belonging to this order that undergoes the ſame metamorphoſis with thoſe of the other orders, all the other apterous inſects being produced in their perfect ſtate, either by the mother, or from the egg. The larva has a forked tail, and ſpins a covering for the pupa, which has feet, of which, however, it can make no ufe, they being immoveable. The larva may be nourifhed in boxes, and fed with flies, of which they are very fond.

They are very fmall, lively, and creep like caterpillars; they pafs fourteen or fifteen days in their larva ftate, before they undergo their fecond changement.

Genus VI. Acarus.

Linn. Syst. Nat. page 1022.

The insects belonging to this genus have eight feet.

Two eyes placed on the sides of the head, remote from one another.

And two articulated tentacula in the form of feet.

Schæffer observes, that the head of the Acarus is united to the thorax, in which it differs from the foregoing genera belonging to this order; that its feet are made for running, its antennæ, (the tentacula of Linnæus) articulated, and made like feet, and that it has a pointed rostrum.

Geoffroy and the last-mentioned author have given to the *Acarus Longicornis, Linn. No.* 29, and another, which Linnæus has since placed among the *Phalangia (Phal. No.* 4, *Cancroides)* the generical name of *Chelifer*; these differ from the other Acari in the form of their antennæ, which are terminated by a kind of claw, resembling that of a crab. They have given the same generical characters to the other Acari, as Linnæus.

The mouth of the Acarus is formed by a very small rostrum enclosed in a sheath; the antennæ are shorter than the proboscis, except in one species, which is called, from that circumstance, the Acarus *Longicornis*. The thorax is of the same size with the head, and so confounded with the abdomen as not to be distinguished but by its hardness. The Acari live chiefly upon other animals, quadrupedes, birds and insects; some of the last-mentioned class are often quite covered with them; others of them live in the water, others upon trees, plants, &c. They are oviparous, but their copulation and metamorphosis have not yet been observed.

Genus VII. Phalangium.

Linn. Syft. Nat. Pag. 1027.

The Phalangium has eight feet.

Two eyes on the fummit of the head, near each other, and two others on the fides.

The antennæ, which are fixed to the fore part of the head, are made like the feet.

The abdomen is round.

The *Phalangium Opilio, Linn. No.* 2, differs from the others in the number of its eyes, which are but two.

According to Schæffer and Geoffroy, the two palpi in this genus are cheliform, and the antennæ formed like feet, and angulated.

The head and thorax are united without any joint.

Only one fpecies of the Phalangium is common in Europe; the feet of this infect are very slender, weak, and liable to be broken. Geoffroy is of opinion, that these feet, when broken, grow again like the claws of a crab, he having once found a specimen with seven entire legs, of the natural, or common length, and the eighth much shorter; he is farther induced to believe it, from the seeming analogy between the Crab and the Phalangium: this matter is curious, and merits observation.

The tarsi are composed of a very great number of short articulations.

The Phalangia are in general nocturnal animals, flying the light, and searching for their prey in the night time; many of them devour the Acari, Wood-lice, spiders, &c. Some of them live in the sea, attached to the bodies of the larger aquatic animals; others live in the trunks of decayed trees. Their manner of copulation and production is wholly unknown.

GENUS VIII. ARANEA, the SPIDER.

LINN. Syft. Nat. Pag. 1030.

The feet of the Araneæ are eight in number.

They have eight eyes.

Their mouth is armed with two crotchets.

Their palpi are two in number, articulated, and headed by the genitalia of the males, in that fex.

The anus contains inftruments for fpinning, fhaped like nipples or teats.

Schæffer adds to the above characters of the Spider, that the feet are made for running, the head united to the thorax, and the abdomen (which is of an oblong oval form) joined to the thorax by a fhort ftalk or pedicle. He has divided this genus into different families, according to the various fituation of the eyes, in which he followed Frifch, Geoffroy, and others. The eyes of fpiders are immoveable, and their ftructure is different from that of the eyes of moft other infects, confifting each of only one lenfe, which deprives them of the faculty of multiplying objects, as their immobility does that of seeing such objects as are placed otherwise than exactly before each eye.

Geoffroy asserts, that all spiders have eight eyes, and that the eye, at each extremity of the line, in the species which Linnæus believed to have only six, is double.

Spiders prey upon all weaker insects, even those of their own species, and are themselves destroyed by Spheges and Ichneumons; they vary in colour according to their age, and often the different sexes of the same species differ in that particular; they cast off or change their skin; they are not preserved perfect in cabinets without great difficulty, on account of their great humidity.

GENUS

Genus IX. Scorpio, the Scorpion.

Linn. Syst. Nat. page 1037.

The Scorpion has eight feet, and two claws, which last are situate on the fore part of the head.

It has eight eyes, three of which are seated on each side of the thorax, and the two others on the back.

The palpi are two in number, and cheliform.

The tail is lengthened out, articulated, and terminated by a sharp, crooked sting.

On the under side, between the breast and abdomen, are placed two instruments, called *pectines*, from their form, which resembles that of a comb.

Linnæus observes, that this genus is not found in Sweden; nor do I know that they are to be met with any where in the northern parts of Europe.

Schæffer adds, to the characters given of Scorpion by Linnæus, that the feet a for running; the head united with the thorax, and the tail long and articulated.

The claws fituate upon the head are, according to the fame author, the antennæ of the infect; Scopoli calls them palpi.

The venom of the Scorpion is accounted more dangerous than that of any other infect, and has been frequently attended with the lofs of life, in hot climates, as we are informed by different travellers.

Genus X. Cancer, the Crab.

Linn. Syst. Nat. page 1038.

The Crab has eight feet (sometimes ten or six) besides two *hands* terminated by the claws.

It has two moveable eyes, generally projecting from the head, or placed upon a kind of stalk.

It has two palpi armed with claws.

The tail is articulated, and unarmed, or without any kind of sting.

This genus is divided into families, as follows:

1. The *Brachyuri, or short tailed crabs,* in which the thorax is either

    Smooth, and the sides of it entire;

    Smooth, with the sides jagged or indented;

    Hairy, or spinous on the upper part;

    Armed

Armed with spines on the upper part; Or

With an uneven surface.

2. *Macrouri, or long tailed crabs*: these are subdivided from the following circumstances:

Those having a smooth thorax;

Those with an uneven or tuberculated thorax;

Those which have the thorax armed with spines;

Those in which the hand is without fingers, and the thorax of an oblong form;

Those in which the shell of the thorax is shorter than that part, which it does not cover entirely.

Some species of each of these families are *parasitici*; these, for the most part, live in the shells of other testaceous animals, and their tails want the leaves, or plates, which terminate the tails of the other crabs.

The

The crab has two long, and two, or four, short antennæ, which last are by some called palpi.

Schæffer observes, that the antennæ of the crabs are long and setaceous (without making mention of the shorter antennæ or palpi) the head united to the thorax, the mouth armed with jaws, and the body covered with a crust or shell.

Geoffroy asserts, that the head of the Cancri *Macrouri* or *long tailed crabs*, is not united with, but distinct from the thorax; the same author numbers the claws among the feet, and calls the shorter antennæ the palpi, as does likewise Scopoli.

The Crabs are long-lived, and change their crustaceous skin every year, which changement is not effected without great difficulty; the instruments of generation are two in number, in each sex, and they copulate breast to breast, in *resupinata femina*; the female carries her eggs, which are exceedingly numerous, in a cluster under her tail. They feed equally upon plants, dead and live animals, and frequently the strong and healthy ones devour such as have just changed their skin, at which time they are weak, languishing,

languishing, and their new skin soft; at this time they likewise fall a prey to many other animals, and chiefly to different species of the marine polypus. Some authors assert, that the Crab changes its stomach and intestines at the same time with the skin.

Genus XI. Monoculus.

Linn. Syft. Nat. page 1057.

The feet of the Monoculus are made for fwimming.

The body is covered with a cruft, or fhell.

The eyes are fixed in the fhell, very near one another.

The infects belonging to this genus have generally been thought to have two eyes, but placed fo near to one another as fcarce to be diftinguifhed; Geoffroy however afferts, that feveral of them have in fact only one eye; to thefe he has preferved the name of *Monoculus.* Thefe are likewife farther diftinguifhed by their antennæ, which in fome are divided and fubdivided into branches, like plants, with feveral lateral hairs, in others are more than two in number. To the remaining ones, in which two eyes are plainly perceptible, he has given the generical name of *Binoculus*; the antennæ in this genus are fetaceous, and the tail forked. The feet, according to the fame author, are fix in number

(in

(in each of his genera;) but, according to Schæffer, they are many in number, and branched. Perhaps that Author miftook the antennæ for feet, and indeed moft of the fpecies make ufe of the antennæ to fwim, and likewife to leap with; he has changed the generical name of our author to that of *Branchipus*. The Monoculi are both oviparous and viviparous; they live in ftagnated waters; fome of them feed upon plants, others attach themfelves firmly to the bodies of different fifh, whofe blood they fuck for their nourifhment; they fwim, or rather fpring upon the water, with great agility; they are in general very fmall, but lay an amazing number of eggs; they lofe all motion, and feem to ceafe to live in fummer, when the great droughts have deprived them of water, but revive when reftored to their proper element.

Linnæus relates, that one fpecies of them, which is of a red colour, is fometimes fo numerous as to make the waters appear as if changed into blood.

Genus XII. Oniscus.

Linn. Syst. Nat. page 1069.

The Oniscus has fourteen feet.

The antennæ are setaceous, and

The body of an oval form.

Geoffroy adds to the above characters of the Oniscus, that the antennæ are bent. He has separated the *Onisc. Aquaticus, Linn. Syst. Nat. No. 11*, from the other species, under the generical name of *Asellus*, on account of the number of antennæ in that insect, which are four; two of these are longer than the others, but they are all bent: he observes that the head, in both these genera, is intimately joined to the thorax.

Schæffer has followed Geoffroy in this division of the Linnæan genus, and observes, that the feet of the Asellus are made for running, that the body is oblong, and the mouth furnished with two palpi.

The Onisci change their skin, like many other apterous insects; it is composed of several crustaceous plates.

They are found frequently in houses, gardens, and woods; some species live in the water; they are sometimes called *Hog-lice*, and one species is made use of in medicine.

Genus XIII. Scolopendra.

Linn. Syst. Nat. page 1062.

The feet, in this genus, are as many in number, on each side, as the segments of the body.

The antennæ are setaceous.

The palpi two in number, and jointed, or formed of various articulations.

The body is depressed, or flat.

Geoffroy and Schæffer assert, that the antennæ of the Scolopendra are filiform, and composed of many short articulations; the feet, according to the same author, are never fewer than twenty-four.

The body of the Scolopendra is flat, and composed of a great many rings, or segments, which augment, as the insect advances in age, till it is fully grown, for which reason the species can rarely be determined with any certainty: it changes its skin in the same manner as the two preceding genera: some species are frequent in gardens, and all humid places, under stones, &c.

Genus XIV. Julus.

Linn. Syst. Nat. page 1036.

The feet, in this genus, are very numerous, being, on each side, twice as many as the segments of the body.

The antennæ are moniliform.

The palpi are two in number, and articulated.

The body is of a semicylindric form.

Geoffroy and Schæffer observe, that the antennæ are composed of five articulations, and the feet always more than an hundred in number.

The Juli differ from the Scolopendræ in the shape of their body, and number of their feet, which last are likewise very short; the skin is exceedingly hard, and is cast off or changed, like that of the Scolopendræ, &c. They are frequent in humid places.

INDEX
OF
SYNONYMOUS GENERA.

ORDER I.
INSECTA COLEOPTERA.

Linnæus		Geoffroy	Schæffer	Scopoli
Genus I. Scarabæus, p. 26		Scarabæus	Idem	Scarabæus
		Copris	Idem	
II. Lucanus 29		Platycerus	Idem	Lucanus
		Dermestes	Idem	Idem
		Boftrichius	Idem	
III. Dermestes 31		Cistela	Idem	Idem
		Silpha	Idem	
		Byrrhus		

INDEX, &c. ORDER I.

Linnæus.		Geoffroy.	Schæffer.	Scopoli.
Genus IV. Ptinus		Ptilinus		
	34	Byrrhus		Bupreſtis
		Bruchus		
V. Hiſter	36	Attelabus	Idem	Hiſter
VI. Gyrinus	38	Idem	Idem	Dytiſcus
VII. Byrrhus	39	Anthrenus	Idem	
VIII. Silpha	40	Dermeſtes	Silpha	Silpha
		Peltis	Idem	
IX. Caſſida	42	Idem	Idem	Idem
X. Coccinella	44	Idem	Idem	Idem
XI. Chryſomela		Galeruca	Idem	Chryſomela
		Chryſomela	Idem	Coccinella
		Cryptocephalus	Idem	Attelabus
		Crioceris	Idem	Bupreſtis
		Diaperis	Idem	
		Altica	Idem	
		Melolontha	Idem	

INSECTA COLEOPTERA.

Linnæus.		Geoffroy.	Scheffer.	Scopoli.
Genus XII. Hiipa				
XIII. Bruchus	50	Crioceris		
	51	Mylabris		Laria
XIV. Curculio	53	{ Curculio	Idem	
		Rhinomacer	Idem	Curculio
		Mylabris	Idem	
XV. Attelabus	57	{ Clerus	Idem	Attelabus
		Rhinomacer	Idem	Curculio
XVI. Cerambyx	60	Prionus	Idem	
		{ Cerambyx	Idem	Cerambyx
		Leptura	Idem	Leptura
		Stenocorus	Idem	
XVII. Leptura	64	{ Leptura	Idem	Leptura
		Cerambyx	Idem	
		Stenocorus	Idem	
XVIII. Necydalis	66	{ Idem	Necydalis	Necydalis
			Mylabris	Cantharis
			Leptura	

266   INDEX, &c.   ORDER I.

		Linnæus.		Geoffroy.	Schæffer.	Scopoli.
Gen.	XIX.	Lampyris	68	{ Lampyris	Idem	
					Pyrochora	Cassida
	XX.	Cantharis	70	{ Cicindela	Telephorus	
				Pyrochroa	Cantharis	Cantharis
					Leptura	
	XXI.	Elater	72	Idem	Idem	Idem
	XXII.	Cicindela	75	Bupreftis	Cicindela	Idem
	XXIII.	Bupreftis	77	Cucujus	Bupreftis	Mordella
	XXIV.	Dyticus	80	{ Dyticus	Idem	Dytifcus
				Hydrophilus	Idem	
	XXV.	Carabus	82	Bupreftis	Carabus	Idem
	XXVI.	Tenebrio	84	Idem	Idem	Idem
	XXVII.	Meloe	87	{ Meloe	Idem	
				Notoxus	Idem	
				Cerocoma	Idem	Meloe
				Cantharis		
	XXVIII.	Mordella	90	Idem	Idem	Idem

INSECTA HEMIPTERA.

Linnæus.		Geoffroy.	Schæffer.	Scopoli.
G. XXIX. Staphilinus	92	Idem	Idem	Idem
XXX. Forficula	94	Idem	Idem	Idem

ORDER II. INSECTA HEMIPTERA.

Linnæus.		Geoffroy.	Schæffer.	Scopoli.
Genus I. Blatta	page 96	Idem	Idem	Idem
II. Mantis	99	Idem	Idem	Gryllus
III. Gryllus	102	{ Gryllus, Acrydium, Locusta	Idem	Gryllus
IV. Fulgora	106	Idem	Idem	
V. Cicada	108	Idem	Idem	Idem
VI. Notonecta	111	{ Notonecta, Corixa	Idem	Notonecta
VII. Nepa	113	{ Hepa, Naucoris	Idem	Nepa

INDEX, &c. ORDER III.

		Linnæus.	Geoffroy.	Scheffer.	Scopoli.
Gen.	VIII.	Cimex	115 Idem	Idem	Idem
	IX.	Aphis	118 Idem	Idem	Idem
	X.	Chermes	120 Psylla	Chermes	Idem
			{ Chermes	Coccus	Idem
	XI.	Coccus	{ Coccus		
	XII.	Thrips	124 Idem	Idem	Idem

ORDER III. LEPIDOPTERA.

		Linnæus.	Geoffroy.	Scheffer.	Scopoli.
Genus	I.	Papilio	127 Idem	Idem.	Idem
	II.	Sphinx	136 Idem	Idem	Idem
	III.	Phalena	139 { Phalena	Phalena	Idem
			{ Pterophorus		
			{ Tinea		

ORDER IV. NEUROPTERA.

	Linnæus.		*Geoffroy.*	*Schæffer.*	*Scopoli.*
Genus I.	Libellula	149	Idem	Idem	Idem
II.	Ephemera		Idem	Idem	Idem
III.	Phryganea	157	{ Perla { Phryganea	Idem Idem	Phryganea Idem
IV.	Hemerobius	162	{ Phryganea { Hemerobius	Idem Phryganea	
V.	Myrmelion	164	Formicaleo	{ Formicaleo { Libelloides	
VI.	Panorpa	166	Idem	Idem	Idem
VII.	Raphidia	168	Idem	Idem	

ORDER V. INSECTA HYMENOPTERA.

	Linnæus,		*Geoffroy.*	*Schæffer.*	*Scopoli.*
Genus I.	Cynips	page 170	{ Cynips { Diplolepis	Cynips	Idem

INDEX, &c.

ORDER VI.

Linnæus.		Geoffroy.	Schæffer.	Scopoli.
Genus II. Tenthredo, pag.	173	{Crabro / Tenthredo	Idem / Idem	Tenthredo
III. Sirex	179	Urocerus	Idem	Ichneumon
IV. Ichneumon	182	{Ichneumon / Eulophus	Ichneumon	Idem
V. Sphex	187	Ichneumon	Sphex	Sphex
VI. Chryfis	191	Vefpa	Chryfis	Sphex
VII. Vefpa	193	Idem	Idem	Idem
VIII. Apis	197	Idem	Idem	Idem
IX. Formica	200	Idem	Idem	Idem
X. Mutilla	203	Idem		Idem.

ORDER VI. INSECTA DIPTERA.

Linnæus.		Geoffroy.	Schæffer.	Scopoli.
Genus I. Oeftrus	page 205	Idem	Idem	
II. Tipula	208	{Tipula / Bibio	Idem / Idem	Tipula

INSECTA DIPTERA.

Linnæus.		Geoffroy.	Schæffer.	Scopoli.
Genus III. Mufca	page 210	Stratyomys	Idem	Mufca
		Mufca	Idem	Ceria
		Vollucella	Idem	Conops
		Nemotelus	Idem	Anthrax
		Scathopfe	Idem	
IV. Tabanus	217	Idem	Idem	Idem
V. Culex	220	Idem	Idem	Idem
VI. Empis	222		Empis	Afilus
VII. Conops	224	Stomoxys	Stomoxoides	{ Empis { Sicus { Rhingia
VIII. Afilus	227	Idem	Idem	{ Erax { Afilus
IX. Bombylius	230	Afilus	Bombylius	Idem
X. Hippobofca	230	Idem	Idem	Idem

271

ORDER VII. INSECTA APTERA.

Linnæus.		page	Geoffroy.	Schæffer.	Scopoli.
Genus I.	Lepisma		Forbicina	Lepisma	Idem
II.	Podura	237	Idem	Idem	Idem
III.	Termes	239	Pediculus	Termes	Idem
IV.	Pediculus	240	Idem	Idem	Idem
V.	Pulex	243	Idem	Idem	Idem
VI.	Acarus	245	{ Chelifer { Acarus	Idem	Acarus
VII.	Phalangium	247	Idem	Idem	Idem
VIII.	Aranea	249	Idem	Idem	Idem
IX.	Scorpio		Idem	Idem	Idem
X.	Cancer		Idem	Idem	Idem
XI.	Monoculus		{ Monoculus { Binoculus	Branchipus	
XII.	Oniscus		{ Oniscus { Asellus	Idem	
XIII.	Scolopendra		Idem	Idem	Idem
XIV.	Julus		Idem	Idem	Idem

www.ingramcontent.com/pod-product-compliance
Lightning Source LLC
Chambersburg PA
CBHW032122230426
43672CB00009B/1830